ARCHITECTURAL ACOUSTICS

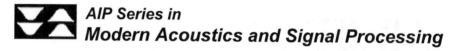

AIP Series in
Modern Acoustics and Signal Processing

ROBERT T. BEYER, Series Editor-in-Chief
Physics Department, Brown University

EDITORIAL BOARD

BOOKS IN SERIES

ARCHITECTURAL ACOUSTICS

Blending Sound Sources, Sound Fields, and Listeners

Yoichi Ando
Kobe University, Japan

With 172 Illustrations

Yoichi Ando
Graduate School of Science
  and Technology
Kobe University
Rokkodai, Nada
Kobe 657-8501
Japan

Series Editor:
Robert T. Beyer
Physics Department
Brown University
Providence, RI 02912
USA

Cover illustrations: The top left photo is by Mitsuo Yamamoto; the bottom right photo is by Hironori Sakimoto.

Library of Congress Cataloging-in-Publication Data
Ando, Yoichi, 1939–
    Architectural acoustics: blending sound sources, sound fields,
    and listeners/Yoichi Ando.
        p. cm. – (Modern acoustics and signal processing)
    Includes bibliographical references and indexes.
    ISBN 0-387-98333-3 (alk. paper)
    1. Architectural acoustics. I. Title. II. Series.
NA2800.A49   1998
690'.2–dc21   97-33261

Printed on acid-free paper.

Production managed by Francine McNeill; manufacturing supervised by Joe Quatela.
Photocomposed copy prepared from the author's electronic files using Springer's sv-pl.sty macro.
Printed and bound by Maple-Vail Book Manufacturing Group, York, PA.
Printed in the United States of America.

9 8 7 6 5 4 3 2 1

ISBN 0-387-98333-3 Springer-Verlag New York Berlin Heidelberg   SPIN 10642723

I gratefully dedicate this volume to my mother.

(August 28, 1906–September 18, 1996)

Series Preface

Soun is noght but air y-broke
—Geoffrey Chaucer
end of the 14th century

Traditionally, acoustics has formed one of the fundamental branches of physics. In the twentieth century, the field has broadened considerably and has become increasingly interdisciplinary. At the present time, specialists in modern acoustics can be encountered not only in physics departments, but also in electrical and mechanical engineering departments, as well as in mathematics, oceanography, and even psychology departments. They work in areas spanning from musical instruments to architecture to problems related to speech perception. Today, six hundred years after Chaucer made his brilliant remark, we recognize that sound and acoustics is a discipline extremely broad in scope, literally covering waves and vibrations in all media at all frequencies and at all intensities.

This series of scientific literature, entitled Modern Acoustics and Signal Processing (MASP), covers all areas of today's acoustics as an interdisciplinary field. It offers scientific monographs, graduate-level textbooks, and reference materials in such areas as architectural acoustics, structural sound and vibration, musical acoustics, noise, bioacoustics, physiological and psychological acoustics, speech, ocean acoustics, underwater sound, and acoustical signal processing.

Acoustics is primarily a matter of communication. Whether it be speech or music, listening spaces or hearing, signaling in sonar or in ultrasonography, we seek to maximize our ability to convey information and, at the same time, to minimize the effects of noise. Signaling has itself given birth to the field of signal processing, the analysis of all received acoustic information or, indeed, all information in any electronic form. With the extreme importance of acoustics for both modern science and industry in mind, AIP Press, now an imprint of Springer-Verlag, initiated this series as a new and promising publishing venture. We hope that this venture will be beneficial to the entire international acoustical community, as represented by the Acoustical Society of America, a founding member of the American Institute of Physics, and other related societies and professional interest groups.

It is our hope that scientists and graduate students will find the books in this series useful in their research, teaching, and studies. As James Russell Lowell once wrote, "In creating, the only hard thing's to begin." This is such a beginning.

Robert T. Beyer
Series Editor-in-Chief

Preface

Concert hall acoustics can be thought of as the place where science and art meet. Imagine how many scientists have contributed to its development. The oldest known writing on the subject, which concerns theater acoustics in ancient Greece and Rome, dates from about 25 B.C. It describes advanced designs for better acoustics, which involved digging holes between chairs and placing bronze vessels upside down in the holes according to mathematics-based music theory. New York Philharmonic Hall, which opened in 1962, was designed on the basis of temporal factors representative of reverberation time. The hall was not well received by the public, however, and was closed in 1978 for extensive renovation. The most important objective is to blend acoustics and music in such a way that each individual's feelings harmonize with the hall's acoustics. Concert hall acoustics has been my major area of study for approximately 30 years; the first 20 years were devoted to the physical and psychological approaches. The goal was to calculate the overall subjective preferences of audience members in each seat. The findings are summarized as follows (Ando, Y., *Concert Hall Acoustics*, Springer-Verlag, Heidelberg, 1985):

(1) A hall should be designed for only a certain type of music, because the sound quality depends on the effective duration of the autocorrelation function of music signals and the temporal acoustic factors of sound fields. Musicians seem to compose their music with acoustics in mind: pipe organ music was written for large spaces, such as Notre Dame Cathedral, and Mozart's string quartets were intended to be heard in court salons.
(2) The newly introduced spatial factor (IACC) is the most effective on subjective preference and subjective diffuseness among acoustic factors of sound fields.
(3) The theory allows calculating the global preference of each seat at the design stage with four physical orthogonal factors of sound fields: the relative sound pressure level (LL), the initial time delay (Δt_1) between the direct sound and the first reflection, the subsequent reverberation time T_{sub}, and the magnitude of the interaural cross-correlation IACC.

For the past 10 years, my focus has been on auditory–brain function and the individual. This research shows that the left cerebral hemisphere is associated with the temporal factors, Δt_1 and T_{sub}, and that the right cerebral hemisphere is activated by the more spatial factors, IACC and LL. The information corresponding to subjective preference of sound fields is found in brain waves. Surprisingly, individual differences in subjective preference appear in brain waves and are recognized mainly in temporal factors and LL, not in IACC. Individual differences in LL may be related to the hearing level. For different values of Δt_1 and T_{sub}, significant individual preferences arise. It is most likely the result of difference in individual temporal activities of the brain. This evidence ensures that the basic theory of subjective preference may be applied to each individual preference as well. The other fundamental subjective attributes for sound fields can also be described by the theory, based on the auditory–brain model with correlation mechanisms and the cerebral-hemispheres specialization.

To blend sound sources and sound fields in a concert hall, the sound fields must first be designed to maximize the average preference at each seat. Musicians must select music programs appropriate for the hall and performing positions on the stage that both maximize the ease of the performers and the preference of the listeners. A seat-selection system for satisfying individual preference was introduced at the Kirishima International Concert Hall in 1994, and the first international symposium of experts in the art and science of sound, Music and Concert Hall Acoustics (MCHA), was held in May 1995. As described in this book, 106 participants took part in a test procedure for seat selection.

I hope that the theory of incorporating temporal and spatial values for both levels of global subjective preferences and individual preference of sound fields can be generalized to blend nature, the built environment, and people.

Yoichi Ando

Acknowledgments

The studies described in this book were performed at the Graduate School of Science and Technology, Kobe University, and partially at the Drittes Physikalisches Institut, University of Göttingen, Germany. I express my appreciation to both institutions and their staffs. Special thanks are due to Professor Manfred R. Schroeder for his continuous invitation to work at his institute since 1971. In the summer of 1995, I started preparing the manuscript for this book in Göttingen at Professor Schroeder's suggestion, as well as at the invitation of Professor Werner Lauterborn, director of the institute.

I am very grateful to Professor Robert T. Beyer, Brown University, for his continuing warm encouragement since my first book, *Concert Hall Acoustics*, was published in 1985 and for the substantial improvement of the English usage in that book. Also, I am deeply grateful to Professor Emeritus Isamu Suda, Kobe University, for his talk in 1979, which provided me with a source of inspiration for research into the work described in Sections 5.2 and 5.3.

Most of the illustrations have previously been published by me and/or my colleagues. I express my appreciation to the authors and publishers who have granted permissions, and I thank Dr. Zvi Ruder and Dr. Charles Doering, then editors at AIP Press, for their encouragement in publishing this book. Finally, the generous support of the Alexander von Humboldt Foundation, Bonn, since 1975, enabled me to concentrate on the research necessary to complete this monograph.

Yoichi Ando

Contents

1

Introduction

A number of investigations have been on-going since my first book, *Concert Hall Acoustics*, was published in 1985. Typical examples are the study of hemispheric brain activities, in order to identify a model of auditory–brain systems, orthogonal physical factors, and the theory of individual subjective preference. This book describes comprehensive concepts, theoretical backgrounds and subjective evaluations, in addition to subjective preference for the sound fields, as well as applications in the design of concert and multiple-purpose halls. Particular emphasis has been placed on enhancing performance in the selection of the "most-preferred" seat for individuals in a hall.

It is also of interest that such a theory may be applied in more general physical environments, such as those of light and heat, taking the spatial and temporal factors into account.

This book is written for both undergraduate and graduate students in various fields including acoustics, psychology and physiology, and musical art, as well as professionals in architecture, engineering, and "sound coordinators" of concert halls and theaters. Readers, who are interested in the applications of designing concert halls and theaters, as well as electroacoustic systems, are recommended to read, first of all, Chapters 10 and 11, and then proceed to the guidelines described in Chapters 4, 6, 7, and 8.

Special attention is given to the process obtaining scientific results, and a model of the auditory–brain system, rather than describing only a final design method. Such processes may therefore help researchers who are interested in fusing science and art to become aware of the appropriate lines for future work.

2

Short Historical Review for Acoustics in a Performing Space

Investigations in architectural acoustics to find dimensional and orthogonal factors influencing the subjective evaluation of sound fields go as far back as Vitruvius (ca. 25 B.C.). In the "ancient architectural acoustics," the concepts of reverberation, interference, echo disturbance, and clarity of voice were described. Vitruvius's remarkable statement that "bronze vessels, which were tuned notes of the fourth, the fifth, and so on," by the calculation due to musical theory, were inverted in niches, and supported on both sides facing the stage by wedges not less than half a foot high. Niches in between the seats of the theater were constructed. Obviously, a great deal of scientific effort was attempted and much attention was devoted at that time in the space under our ears. In this book, the acoustic design of the floor structure and seating will also be discussed.

Because of the lack of electroacoustic techniques, far fewer investigations on architectural acoustics than the ancient ones were reported between the first and nineteenth centuries.

In 1857, Henry first mentioned the concept of the impulse, which is utilized in the modern science. In his case, a single impulse from one tooth of a wheel is a noise, from a series of teeth in succession a continued sound; and if all the teeth are equally spaced, and the speed of the wheel is uniform, then a musical note is the result. Further, he suggested factors that might be related to good acoustics indicating the following conditions:

(1) the size of the room;
(2) the strength of the sound or intensity of the impulse;
(3) the position of the reflecting surfaces; and
(4) the nature of the material of the reflecting surfaces.

It is interesting to note that these conditions are, to some extent, related to the four orthogonal factors that are described in this book.

Sabine (1900) initiated the science of architectural acoustics, discovering reverberant sound, and the formula to quantify reverberation time. The first careful experiment on the absolute rate of decay was in the lecture-room of the Boston Public Library, a large room. On the platform four organ pipes were placed, all of the same pitch, each with its own wind supply, and each having its own electropneumatic value. Thus, one, two, three, or four pipes might start and stop at once. Then, the minimum audible times were measured. The corresponding durations of audibility, named t_1, t_2, t_3, and t_4 are: 8.68 s, 9.14 s, 9.36 s, and 9.55 s, respectively. The time differences were obtained as follows:

$$t_2 - t_1 = 0.45\,[\text{s}],$$
$$t_3 - t_1 = 0.67\,[\text{s}], \qquad\qquad (2.1)$$
$$t_4 - t_1 = 0.86\,[\text{s}].$$

The exponential decay rate of the intensity can be obtained by the use of these time differences. Consequently, he finally derived the well-known reverberation formula

$$T_{60} = KV/A, \qquad\qquad (2.2)$$

where K is a constant ($= 0.159$) for the velocity of sound at 342 m/s, V is the volume of the room in cubic meters, and A is the absorbing power of the room.

Sabine recognized the Chapel of the Union Theological Seminary, New York City, as a very satisfactory example, without any explanation (Sabine, 1912). Considering the fact that the shape of the ceiling in the chapel is like the bottom of a boat, which effectively decreases the IACC as described in Section 8.2, it seems that Sabine unconsciously noticed the importance of the spatial shape of a room.

Knudsen (1929) suggested that the optimal reverberation time for speech is shorter than that for music. At the same time, MacNair (1930) recommended a longer reverberation time in the low-frequency range to supplement the loudness of music.

Békésy (1934) reported that a courtyard sound field, as shown in Figure 2.1, was much better than any of the sound fields in existing halls he had experienced. This clearly suggested the importance of side-wall reflections (Figure 7.9).

In 1949, Haas investigated the echo disturbance effects by adjustment of the delay time of the early reflection by moving the head-positions of a magnetic tape recorder. He showed the disturbance of speech echo to be a function of the delay time and, as a parameter of the amplitude, Bolt and Doak (1950) later proposed the percent disturbance of echoes.

Considering the fact that living creatures emerged and evolved in physical environments, including acoustic, visual, and thermal environments, our sensing organs and brain are greatly influenced by the physical environmental factors that existed before emergence. In Table 2.1, since 1960, physical factors found by several authors, which significantly influence subjective attributes, are listed. After investigation of a number of existing concert halls throughout the world, Beranek (1962) proposed a rating scale with eight factors of the sound field, from data obtained by questionnaire on existing halls, given to experienced listeners. Much attention has been given to the temporal factors of the sound field since Sabine's discovery of reverberation theory. Clearly, the binaural effect was not satisfactory to listeners.

Venekalasen and Christoff (1964) suggested the importance of reflections from the side walls. West (1966) found the correlation coefficient for $2H/W$ (H is

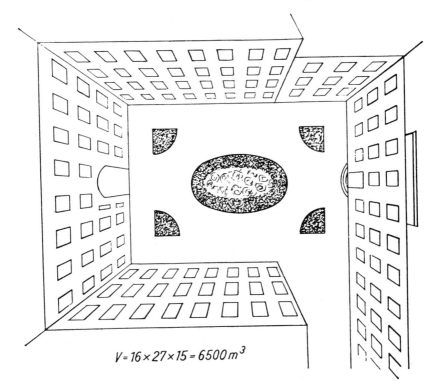

$$V = 16 \times 27 \times 15 = 6500\, m^3$$

FIGURE 2.1. A courtyard with superior acoustics (Békésy, 1934; Békésy, 1967).

TABLE 2.1. Significant physical factors found for sound fields, both existing and simulated, from systematically subjective judgment tests.

Author	Year	Subjective judgment	Sound system	Number of factors	Objective and subjective significant factors found and/or proposed
Beranek*	1962	Questionnaire	Listening in existing halls	8[6][†]	(1) Initial time delay; (2) Loudness; (3) Reverberation time; RT; (4) Frequency characteristic of RT; and others.
Keet	1968	Apparent source width	Simulated	2	(1) SPL; (2) ICC.[‡]

TABLE 2.1.   *cont.*

Author	Year	Subjective judgment	Sound system	Number of factors	Objective and subjective significant factors found and/or proposed
Barron	1971	Spatial impression	Direct sound and early reflections simulated	2	(1) SCC; (2) SPL; (3) Spectrum.
Damaske and Ando	1972	Subjective diffuseness and direction of sound source	Dummy head and loudspeaker	2	(1) IACC; (2) τ_{IACC}.
Yamaguchi	1972	Dissimilarity	Two microphones and headphones	3[2]	(1) SPL; (2) Frequency characteristic of RT.
Edward	1974	Dissimilarity	Dummy head and headphones	3	(1) RT; (2) Volume level; (3) Early echo pattern.
Schroeder, Gottlob, and Siebrasse	1974	Preference by paired comparison	Dummy head and loudspeakers	4[2]	(1) RT; (2) IACC.
Ando	1977	Preference by paired comparison	Loudspeakers simulation for direct sound and the first reflection	2	(1) Initial time delay; (2) IACC.
Ando	1983	Preference by paired comparison	Loudspeaker simulation	4	(1) Listening level; (2) Initial time delay; (3) Subsequent reverberation time; (4) IACC.
Cocchi, Farina, and Rocco	1990	Preference	Real sound field in a hall	4	(1) Listening level; (2) Initial time delay; (3) Subsequent reverberation time; (4) IACC.
Sato, Mori, and Ando	1997	Preference by paired comparison	Real sound field switching source positions (loudspeakers) at fixed seats in a hall	4	(1) Listening level; (2) Initial time delay; (3) IACC; (4) τ_{IACC}.

[*] Numbers in brackets indicate the numbers of dimensions that may be regarded as significant factors.

[†] Beranek (1996) later proposed six factors, but two added factors are questionable in their orthogonality. With regard to the frequency characteristics of the reverberation time, the range below 500 Hz is not critical in preference judgments, so that the preferred range is broad (Ando, Okano, and Takezoe, 1989).

[‡] Short-term cross-correlation coefficient.

the height and W is the width of a hall) and a numerical scale of subjective categories to be 0.71. Damaske (1967/68) investigated subjective diffuseness by arranging a number of loudspeakers around the listener. Keet (1968) reported the variation of apparent source width (ASW) in relation to the interaural cross-correlation coefficient and the sound pressure level. Marshall (1968a,b) stressed the importance of early lateral reflections of just 90°, and Barron (1971) investigated "spatial impressions" or "envelopement" of sound fields in relation to the interaural cross-correlation coefficient. Damaske and Ando (1972) defined the IACC as the maximum absolute value of the interaural cross-correlation function within the possible maximum interaural delay range for humans such that

$$\text{IACC} = |\phi_{lr}(\tau)|_{\max} \quad \text{for} \quad |\tau| \leq 1 \text{ ms,} \tag{2.3}$$

and proposed the method of calculating the interaural cross-correlation function for the sound fields.

By dissimilarity tests, Yamaguchi (1972) reported that the sound-pressure level and the frequency characteristics are significant factors for a sound field recorded in a hall. Edward (1974) also tested the dissimilarity of recorded sound fields, and reported as important factors the early-echo pattern as well as RT and volume level. Schroeder, Gottlob, and Siebrasse (1974) reported results of the paired-comparison tests asking which of two sounds of listened music were preferred. Sound fields were reproduced at each ear of a listener in an anechoic chamber, through dummy head recording and two loudspeaker systems with filters reproducing spatial information. They found two significant factors, RT and IACC, having a strong influence on subjective preference. Wilkens (1977) claimed that significant subjective attributes were perception of strength and extension of sound source, as well as perception of clarity and tone color.

Ando and Kageyama investigated subjective preference in relation to physical factors, which were calculated from the mathematical expression for sound arriving at both ears (Ando, 1977; Ando and Kageyama, 1977). In 1983, Ando published a theory of subjective preference in relation to the four orthogonal physical factors for a sound field, enabling the calculation of a scale value at each seat (see also Ando, 1985, 1986). This theory was first confirmed by Cocchi, Farina and Rocco (1990) in an existing hall. Sato, Mori, and Ando (1997) reconfirmed it more clearly by the paired-comparison judgments in an existing hall, switching the loudspeakers on the stage instead of changing seats. They introduced the interaural delay of the IACC, τ_{IACC}, for the image shift of the sound source that is to be avoided or for the balance of the sound field.

Thus far, this theory has been based on the global subjective attributes for a number of subjects. In order to further enhance each individual's satisfaction, the theory may be applied by adjusting the weighting coefficient of each orthogonal factor (Ando and Singh, 1996; Singh and Ando, unpublished), even though a certain amount of inter-individual differences exist (Sakai, Singh, and Ando, 1997). The seat selection system (Sakurai, Korenaga, and Ando, 1997), which was introduced after construction of the Kirishima International Concert Hall, is a typical example of this application (Ando and Setoguchi, 1995).

3

Physical Properties of Source Signals and Sound Fields in a Room

Sound signals proceed along auditory pathways and are perceived in a time sequence, and the meanings of the signals are simultaneously interpreted by the brain. Thus, a great deal of attention is paid here to analyzing signals in the time domain. This chapter treats mainly the autocorrelation function (ACF) of the signal, which contains the envelop and its fine structure as well as the power at the starting time. The ACF has the same information as the power density spectrum of the signal under analysis, but the ACF differs greatly from the spectrum insofar as the signal processing in the auditory–brain system and the related subjective attributes for the sound field are concerned.

3.1. Analyses of Source Signals

3.1.1. Power Density Spectrum

Let us first discuss signal analysis in the frequency domain in terms of the power density spectrum of a signal $p(t)$, which is defined by

$$P_d(\omega) = P(\omega)P^*(\omega), \qquad (3.1)$$

where $P(\omega)$ is the Fourier Transform of $p(t)$, given by

$$P(\omega) = \frac{1}{2\pi} \int_{-\infty}^{+\infty} p(t)e^{-j\omega t}\, dt. \qquad (3.2)$$

and the asterisk denotes the conjugate.

The inverse Fourier Transform is the original signal $p(t)$:

$$p(t) = \int_{-\infty}^{+\infty} P(\omega)e^{j\omega t}\, d\omega. \qquad (3.3)$$

In considering the sharpening effects existing in the auditory system, the required sharpness of filters used in hearing tests must have slope characteristics of more than 1000 dB/octave. This will be examined under the Results of loudness in Section 6.3.

3.1.2. Long-Time Autocorrelation Function (ACF) of a Sound Source

One of the most promising signal processes in the auditory system is the ACF, which is defined by

$$\Phi_p(\tau) = \lim_{T \to \infty} \frac{1}{2T} \int_{-T}^{+T} p'(t)p'(t + \tau)\, dt, \tag{3.4}$$

where $p'(t) = p(t) * s(t)$, $s(t)$ being the ear sensitivity. For practical convenience, $s(t)$ may be chosen as the impulse response of an A-weighted network. Also, the ACF can be obtained from the power density spectrum, which defined by Equation (3.1), so that

$$\Phi_p(\tau) = \int_{-\infty}^{+\infty} P_d(\omega)e^{j\omega t}\, d\omega, \tag{3.5}$$

$$P_d(\omega) = \int_{-\infty}^{+\infty} \Phi_d(\tau)e^{-j\omega\tau}\, d\tau. \tag{3.6}$$

Thus, the ACF and the power density spectrum mathematically contain the same information. In the ACF analysis, there are three significant parameters, namely,

(1) the energy represented at the origin of the delay $\Phi_p(0)$;
(2) the effective duration of the envelope of the normalized ACF, τ_e, which is defined by the ten-percentile delay, representing a kind of repetitive feature or reverberation contained within the source signal itself; and
(3) the fine structure, including peaks with its delays and the zero crossing number.

The normalized ACF is defined by

$$\phi_p(\tau) = \frac{\Phi_p(\tau)}{\Phi_p(0)}. \tag{3.7}$$

Examples of analyzing the normalized ACF ($2T = 35$ s) for the two extreme music motifs listed in Table 3.1 are shown in Figure 3.1.

When $p'(t)$ is measured with reference to the pressure 20 μPa leading to the level $L(t)$, the equivalent sound pressure level L_{eq}, defined by

$$L_{eq} = 10 \log \frac{1}{T} \int_0^T 10^{\frac{L(t)}{10}}\, dt, \tag{3.8}$$

corresponds to

$$10 \log \Phi_p(0). \tag{3.9}$$

This is an important factor related to loudness, but it is not the whole story. The envelope of the normalized ACF is also related to important subjective attributes, as will be detailed in the subsequent chapters.

A good example of applying the ACF is in the discussion of the missing fundamentals of music signals. When the signal contains only a number of harmonics without the fundamental frequency, we hear the fundamental frequency as a pitch

TABLE 3.1.   Music and speech source signals and their effective duration of the ACF, τ_e.

Sound source*	Title	Composer or writer	τ_e[†] [ms]		$(\tau_e)_{min}$[‡]
Music Motif A	Royal Pavane	Orlando Gibbons	127	(127)	125
Music Motif B	Sinfonietta, Opus 48; IV movement	Malcolm Arnold	43	(35)	40
Music Motif B(L + R)	Sinfornietta, Opus 48; IV movement	Malcolm Arnold			45
Music Motif C	Symphony No. 102 in B flat major; II movement	Franz J. Haydn		(65)	
Music Motif C(L + R)	Symphony No. 102 in B flat major; II movement	Franz J. Haydn			70
Music Motif D	Siegfried Idyll; Bar 322	Richard Wagner		(40)	
Music Motif E	Symphony in C major, K-V, no. 551, IV movement	Wolfgang A. Mozart	38		
Music Motif F	§	Tsuneko Okamoto	105		
Music Motif G	§	Tsuneko Okamoto	145		
Music Motif K	Karesansui	Hozan Yamamoto	220		35
Speech S	Poem read by a female	Doppo Kunikida	10	(12)	

* The left channel signals of the original recorded signals (Burd, 1969) were used, and (L + R) indicates the accompanying right channel signal was mixed with the left channel signal containing the main melody.

[†] Values of τ_e differ slightly with different radiation characteristics of the loudspeakers used; thus all of the physical factors must be measured at the conditions of the hearing tests, $2T = 35$ s.

[‡] Recommended method: the value of $(\tau_e)_{min}$ is obtained by the minimum value of short-moving ACFs, $2T = 2$ s, with the moving interval of 100 ms;

§ Composed for preference judgments of alto-recorder soloists (Section 7.1).

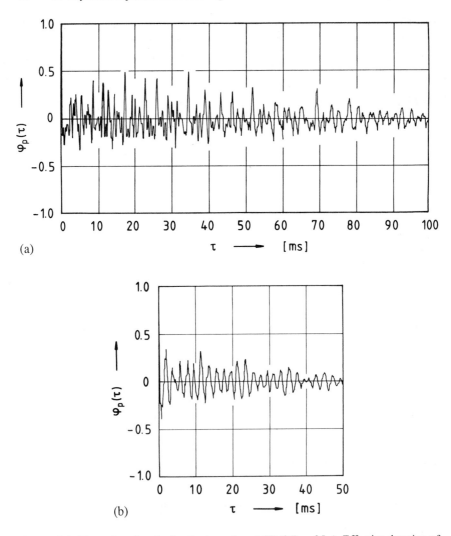

FIGURE 3.1. Examples of analyzing the long-time ACF ($2T = 35$ s). Effective duration of the normalized ACF is defined by the delay τ_e at which the envelope of the normalized ACF becomes 0.1. (a) Music Motif A; Royal Pavane, composed by Gibbons, $\tau_e = 127$ ms; and (b) Music Motif B: Sinfonietta, Opus 48, IV movement, Allegro con brio, $\tau_e = 43$ ms. Note that, according to the characteristics of the loudspeaker used in the subjective judgments, the effective duration of the ACF may differ slightly, for example, $\tau_e = 35$ ms (Music Motif B).

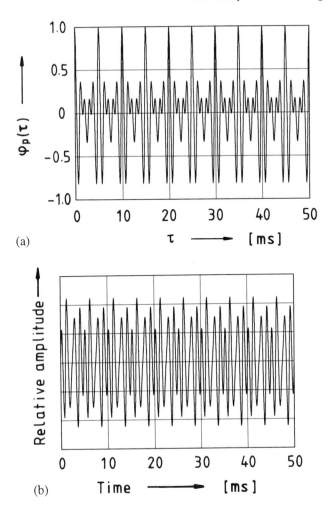

FIGURE 3.2. (a) The normalized ACF of the harmonic components of $3 f_0$, $4 f_0$ and $5 f_0$ in which the missing fundamental (perceived pitch) is $f_0 = 200$ Hz; and (b) the real wave form has three harmonic components with the third harmonic of out-phase.

(Wightman, 1973). This phenomenon is well explained by the ACF-fine structure shown in Figure 3.2 (Sumioka and Ando, 1996). The normalized ACF of only third, fourth, and fifth harmonics clearly contains the period of the fundamental frequency in the fine structure of the ACF as shown in Figure 3.2(a), but is not clear in the real sound signal in time as shown in Figure 3.2(b). In the auditory pathways, therefore, the sound signals are assumed to be processed by the ACF in the time domain. This cannot be explained by the spectrum in the frequency domain only (see also Shoda and Ando, 1996).

Other important subjective attributes of a sound field are best described based on the ACF of the source signals, as detailed in Chapters 4 through 7.

3.1.3. Short-Time Moving Autocorrelation Function of a Source Signal

It is interesting to consider the fact that, in producing a sound signal from any kind of musical instrument, the radiated sound comes from a nonlinear process. Therefore, it may produce special and various sound properties that seem to be known and utilized by many musicians in their performance, without any knowledge of acoustic science. This includes the phenomena of bifurcations and chaos, yielding, for instance, quasi-periodic oscillations that may come from a feedback with a time delay that occurs in most musical instruments (Gueth, 1987; Lauterborn and Parlitz, 1988). In the case of string instruments, the oscillation comes from coupling of the string with the resonance of the bridge in the time domain (Müller and Lauterborn, 1996). Effects of such properties have been tackled by the method of analyzing the time series or the phase-space representation.

Since a certain degree of coherence exists in the time sequence of the source signals, which may greatly influence subjective attributes of the sound field, use is made here of the short ACF as well as the long-time ACF.

The short-time moving ACF as a function of time t is calculated as

$$\phi_p(\tau) = \phi_p(\tau; t, T)$$
$$= \frac{\Phi_p(\tau; t, T)}{[\Phi_p(0; t, T)\Phi_p(0; \tau + t, T)]^{1/2}}, \quad (3.10)$$

where

$$\Phi_p(\tau; t, T) = \frac{1}{2T} \int_{t-T}^{t+T} p'(s)p'(s + \tau)\,ds. \quad (3.11)$$

The normalized ACF satisfies the condition that $\phi_p(0) = 1$.

In order to demonstrate a procedure obtaining the effective duration of the short-time ACF analyzed, Figure 3.3 shows the absolute value in the logarithmic form as a function of the delay time. The envelope decay of the initial and important part of ACF may be fitted by a straight line in most cases. The effective duration of ACF, defined by the delay τ_e at which the envelope of the ACF becomes -10 dB (or 0.1; the ten percentile delay), can easily be obtained by the decay rate extrapolated in the range from 0 dB, at the origin, to -5 dB.

The short-time effective durations of the ACF for various signal duration $2T$ s with the moving interval are obtained in such a way. Examples of analyzing the moving ACF of the music motif K are shown in Figure 3.4(a) through (f). The signal duration corresponding to the psychological present, as suggested by Fraisse (1982), is $2T = 0.5$–5.0 s. Figure 3.5(a) through (f) shows the moving τ_e for music motifs A and B, $2T = 2.0$ s and 5.0 s. The psychological present defined here is a short time duration of stimuli needed for subjective responses. Since the minimum value of the moving τ_e is the most active part of each piece, containing

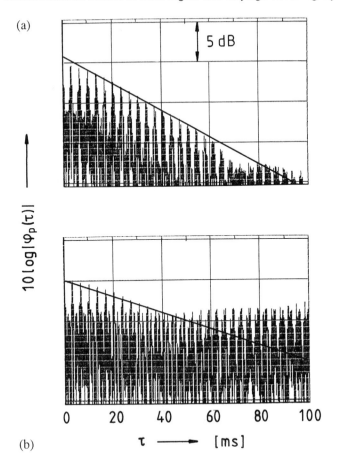

FIGURE 3.3. Examples of determining the effective duration of the running ACF (Music Motif K). (a) $\tau_e = 65$ ms; and (b) $\tau_e = 100$ ms.

important information and influencing subjective responses for the temporal criteria as discussed in Chapter 6 (Ando, Okana, and Takezoe, 1989; see also Mouri and Ando, 1998), the values of $(\tau_e)_{min}$ are plotted in Figure 3.6 as a function of $2T$.

It is interesting that stable values of $(\tau_e)_{min}$ may be obtained in the range of $2T = 0.5$ to 2.0 s for these extreme music motifs.

3.2. Autocorrelation Function of Piano Signal with Varying Performing Style

One of the typical music sources, the piano signal, is selected here. In order to examine the behavior of the ACF of piano signals of varying performing styles, a piano was controlled by a computer for reproduction of the source signal, and

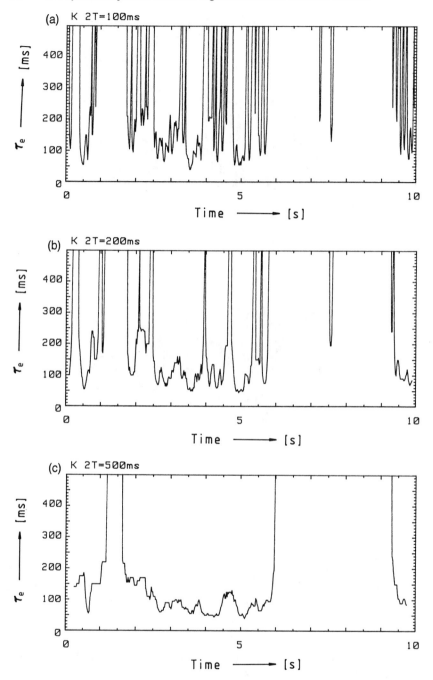

FIGURE 3.4A, B, C. Effective duration of the running ACF with a 100 ms interval as a function of the integration time, $2T$, of Music Motif K (10 s). (a) $2T = 100$ ms; (b) $2T = 200$ ms; and (c) $2T = 500$ ms.

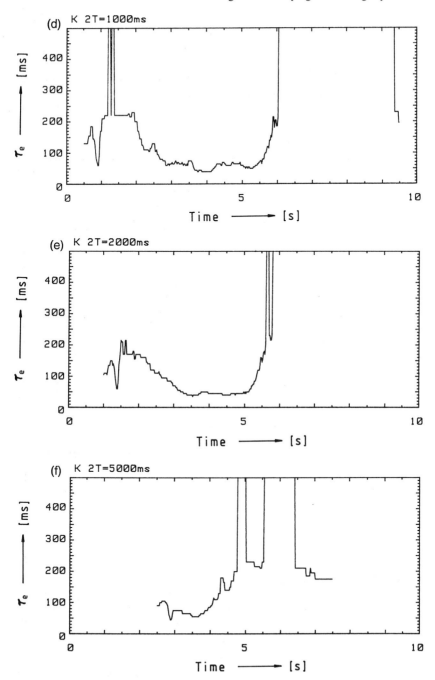

FIGURE 3.4D, E, F. Effective duration of the running ACF with a 100 ms interval as a function of the integration time, $2T$, of Music Motif K (10 s). (d) $2T = 1$ s; (e) $2T = 2$ s; and (f) $2T = 5$ s.

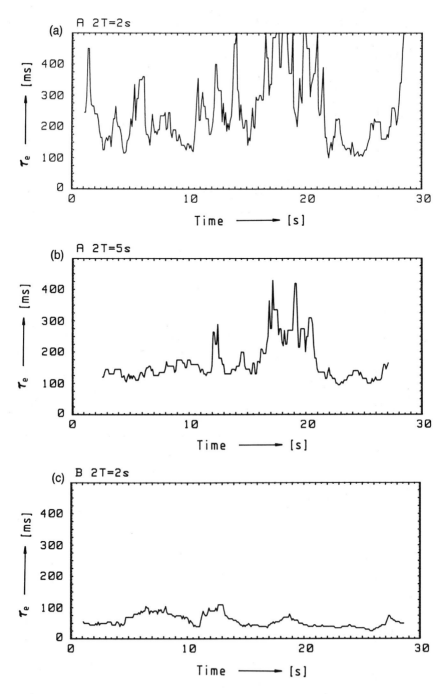

FIGURE 3.5A, B, C. Effective duration of the running ACF with a 100 ms interval as a function of the integration time, $2T$, of the source signals (30 s). (a) Music Motif A (Gibbons), $2T = 2$ s; (b) Music Motif A (Gibbons), $2T = 5$ s; and (c) Music Motif B (Arnold), $2T = 2$ s.

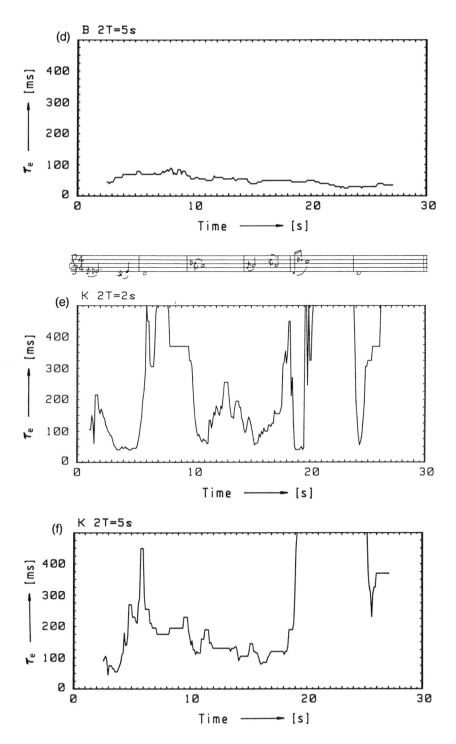

FIGURE 3.5D, E, F. Effective duration of the running ACF with a 100 ms interval as a function of the integration time, $2T$, of the source signals (30 s). (d) Music Motif B (Arnold), $2T = 5$ s; (e) Music motif K (Yamamoto), $2T = 2$ s; and (f) Music Motif K (Yamamoto), $2T = 5$ s.

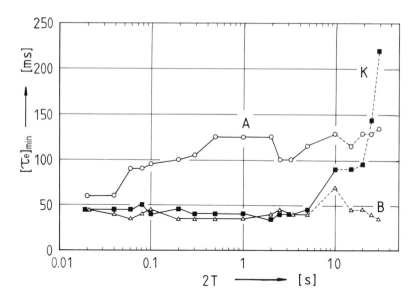

FIGURE 3.6. Minimum values of the effective duration of the ACF as a function of $2T$. (○): Music Motif A (Gibbons); (△): Music Motif B (Arnold); and (■): Music Motif K (Yamamoto).

signals recorded in an anechoic chamber were analyzed (Taguti and Ando, 1997). As is described in Chapters 4 and 6, the effective duration of the ACF, τ_e, is the fundamental time unit of the sound field in a concert hall (e.g., see Figure 6.11). Of particular importance is how the played music fuses with the total sound field. The performance style is basically related to important subjective attributes, such as temporal factors of the sound field, and determines the most preferred initial time delay gap between the direct sound and the first reflection, and also the optimum subsequent reverberation time. If the effective duration of the ACF is varied by the performing style, then the musician may control it to approach the preferred temporal condition of the sound fields, for both performer and listeners to fit in with the performing music and the sound field.

In piano performances, the effective duration of the ACF may be controlled by means of:

(1) speed of performance or tempo;
(2) dynamics;
(3) articulation;
(4) synchronization;
(5) pedaling; and
(6) note-off tail.

Typical examples of the ACF for changing styles of piano performance—staccato and legato—are shown in Table 3.2. As expected, a fast tempo results

TABLE 3.2.   Various styles of piano performance and the effective
duration of the ACF, τ_e. The music piece used is the opening eight bars
of Exercise No. 1, Hanon Tempo, mm $= 120$ under constant dynamics.

Style of performance	NOD [ms]	s [%]*	τ_e [ms]
Staccato	50	70	61–87
Legato	125	0	106–170
Super legato	160	−30	170–233
Mixed	—	—	110–155

* $s = (\text{IOI} - \text{NOD})/\text{IOI}$, where IOI is the inter-onset interval and NOD is the
note-on duration.

in a short value of the effective duration of the ACF, τ_e; and a slow tempo leads
to a long value. The use of the damper pedal creates long values of the τ_e. The
minimum values of τ_e correspond roughly to values of the note-onset duration
(NOD) of a note as shown in Table 3.2. Staccato shortens the values of τ_e as the
acuteness increases, but the values become no shorter than the minimum value
of 60 ms. This lower limit may be caused by a mechanism in producing sound
from the piano. So far, we have noted that the ACF, τ_e, of source signals may be
controlled by changing the performing style.

3.3. Sound Transmission from a Point Source to Binaural Entrances

Let us consider the sound transmission from a source point in a free field to the
binaural earcanal entrances. Let $p(t)$ be the source signal as a function of time, t,
and let $g_l(t)$ and $g_r(t)$ be impulse responses between the source point r_0 and the
binaural entrances. Then the sound signals arriving at the entrances are expressed
by

$$f_l(t) = p(t) * g_l(t),$$
$$f_r(t) = p(t) * g_r(t),$$

(3.12)

where the asterisk denotes convolution.

The impulse responses $g_{l,r}(t)$ include the direct sound and reflections $w_n(t - \Delta t_n)$ in the room as well as the head-related impulse responses $h_{nl,r}(t)$, such that

$$g_{l,r}(t) = \sum_{n=0}^{\infty} A_n w_n(t - \Delta t_n) * h_{nl,r}(t),$$

(3.13)

where n denotes the number of reflections with horizontal angle ξ_n and elevation
η_n, $n = 0$ signifies the direct sound ($\xi_0 = 0$, $\eta_0 = 0$); $A_0 w_0(t - \Delta t_0) = \delta(t)$,

$\Delta t_0 = 0$, $A_0 = 1$, $\delta(t)$ being the Dirac delta function, and A_n is the pressure amplitude of the nth reflection $n > 0$; $w_n(t)$ is the impulse response of the walls for each path of reflection arriving at the listener, Δt_n being the delay time of reflection relative to that of the direct sound, and $h_{nl,r}(t)$ are the impulse responses for diffraction of the head and pinnae for the single sound direction of n. Therefore, Equation (3.12) becomes

$$f_{l,r}(t) = \sum_{n=0}^{\infty} p(t) * A_n w_n(t - \Delta t_n) * h_{nl,r}(t). \tag{3.14}$$

If the source has a certain directivity, $p(t)$ is replaced by $p_n(t)$.

3.4. Physical Factors of Sound Field

3.4.1. Temporal-Monaural Criteria

As far as the auditory system is concerned, all factors influencing any subjective attributes must be included in the sound pressures at the binaural entrances, these are expressed by Equation (3.14).

The first important parameter which depends on the source program is the sound signal $p(t)$. This is represented by the ACF defined by Equation (3.4). The ACF is factored into the energy of the sound signal $\Phi_p(0)$ and the normalized ACF as expressed by Equations (3.4) through (3.7). The normalized ACF includes its envelope, represented by τ_e, peak-amplitudes with the delays, and the zero-crossing number monaural criterion.

The second parameter is the set of impulse responses of the reflecting walls, $A_n w_n(t - \Delta t_n)$. The amplitudes of reflection relative to that of the direct sound $A_1, A_2, \ldots$ are determined by the pressure decay due to the paths d_n, such that

$$A_n = \frac{d_0}{d_n}, \tag{3.15}$$

where d_0 is the distance between the source point and the center of the listener's head. The impulse responses of reflections to the listener $w_n(t - \Delta t_n)$, with delay times of $\Delta t_1, \Delta t_2, \ldots$ relative to that of the direct sound, is given by

$$\Delta t_n = \frac{d_n - d_0}{c}. \tag{3.16}$$

These parameters are not physically independent; in fact, the values of A_n are closely related to Δt_n in such a manner that

$$\Delta t_n = \frac{d_0(1/A_n - 1)}{c}. \tag{3.17}$$

In addition, the initial time-delay gap between the direct sound and the first reflection Δt_1 is statistically related to $\Delta t_2, \Delta t_3, \ldots$, which depend on the dimensions of the room. In fact the echo density is proportional to the square of the time delay

(Kuttruff, 1991). Thus, the initial time-delay gap Δt_1 is regarded as a representation of both sets of Δt_n and A_n ($n = 1, 2, \ldots$).

Another parameter is the set of the impulse responses of the nth reflection, $w_n(t)$ being expressed by

$$w_n(t) = w_n(t)^{(1)} * w_n(t)^{(2)} * \cdots * w_n(t)^{(i)}, \tag{3.18}$$

where $w_n(t)^{(i)}$ is the impulse response of the ith wall existing in the path of the nth reflection from the source to the listener.

Such a set of impulse responses $w_n(t)^{(i)}$ may be represented by a statistical decay rate, namely the subsequent reverberation time, T_{sub}, because $w_n(t)^{(i)}$ includes the absorption coefficient as a function of frequency. This coefficient is given by

$$\alpha_n(\omega)^{(i)} = 1 - |W_n(\omega)^{(i)}|^2. \tag{3.19}$$

According to Sabine's formula (1900), the subsequent reverberation time is approximately calculated by

$$T_{sub} \approx \frac{KV}{\bar{\alpha}S}, \tag{3.20}$$

where K is a constant (about 0.162), V is the volume of the room, S is the total surface, $\bar{\alpha}$ is the average absorption coefficient of the walls, and $\bar{\alpha}S$ is given by the summation of the absorption of each surface i, so that

$$\bar{\alpha}S = \sum_i \alpha(\omega)^{(i)} S^{(i)}. \tag{3.21}$$

3.4.2. Spatial-Binaural Criteria

Two sets of head-related impulse responses for the two ears $h_{nl,r}(t)$ constitute the remaining objective parameter. These two responses $h_{nl}(t)$ and $h_{nr}(t)$ play an important role in sound localization and spatial impression, but are not mutually independent objective factors. For example, $h_{nl}(t) \sim h_{nr}(t)$ in the median plane ($\xi = 0°$) and there are certain relations between them for any other directions to a listener. In fact, a certain relationship between the IATD and the IALD can be expressed for a single directional sound arriving at a listener for a given source signal, and thus for any sound field with multiple reflections. A particular example is that, when the IATD is zero, then the IALD is nearly zero.

Therefore, to represent the interdependence between two impulse responses, a single factor may be introduced, i.e., the interaural cross-correlation function between the sound signals at both ears $f_l(t)$ and $f_r(t)$, which is defined by

$$\Phi_{lr}(\tau) = \lim_{T \to \infty} \frac{1}{2T} \int_{-T}^{+T} f_l'(1) f_r'(t + \tau)\, dt, \qquad |\tau| \leq 1 \text{ ms}, \tag{3.22}$$

where $f_l'(t)$ and $f_r'(t)$ are approximately obtained by signals $f_{l,r}(t)$ after passing through the A-weighted network, which corresponds to the ear sensitivity, $s(t)$. The ear sensitivity may be characterized by the external and the middle ear as described in Section 5.1.

The normalized interaural cross-correlation function is defined by

$$\phi_{lr}(\tau) = \frac{\Phi_{lr}(\tau)}{\sqrt{\Phi_{ll}(0)\Phi_{rr}(0)}}, \tag{3.23}$$

where $\Phi_{ll}(0)$ and $\Phi_{rr}(0)$ are autocorrelation functions at $\tau = 0$ for the left and right ear, respectively, or the sound energies arriving at both ears.

Also, the denominator of Equation (3.23),

$$\sqrt{\Phi_{ll}(0)\Phi_{rr}(0)} \tag{3.24}$$

is the geometrical mean of the sound energies arriving at the two ears.

If discrete reflections arrive after the direct sound, then the normalized interaural cross-correlation is expressed by

$$\phi_{lr}^{(N)}(\tau) = \frac{\sum_{n=0}^{N} A^2 \Phi_{lr}^{(n)}(\tau)}{\sqrt{\sum_{n-0}^{N} A^2 \Phi_{ll}^{(n)}(0) \sum_{n=0}^{N} A^2 \Phi_{rr}^{(n)}(0)}}, \tag{3.25}$$

where we put $w_n(t) = \delta(t)$, and $\Phi_{lr}^{(n)}(\tau)$ is the interaural cross-correlation of the nth reflection, $\Phi_{ll}^{(n)}$ and $\Phi_{rr}^{(n)}(0)$ are the respective sound energies arriving at the two ears from the nth reflection. The denominator of Equation (3.25) indicates the geometric mean of the sound energies at the two ears.

The magnitude of the interaural cross-correlation is defined by

$$\text{IACC} = |\phi_{lr}(\tau)|_{\max} \tag{3.26}$$

for the possible maximum interaural time delay, say,

$$|\tau| \leq 1 \text{ ms.}$$

When the sound source is located at any horizontal angle ξ relative to the frontal direction to a listener's head, and the bandpass noise, after passing through an ideal filter with upper and lower frequencies of f_2 and f_1, is radiated from the source location, then the interaural cross-correlation function and the autocorrelation function at $\tau = 0$ are given by

$$\Phi_{lr}(\tau) = H_{lr} \left[\frac{2}{\Delta\omega(\tau - \tau_\xi)} \right] \sin \left[\frac{\Delta\omega(\tau - \tau_\xi)}{2} \right] \cos \left[\frac{\Delta\omega_c(\tau - \tau_\xi)}{2} \right],$$
$$\Phi_{ll}(0) = H_{ll}, \tag{3.27}$$
$$\Phi_{rr}(0) = H_{rr},$$

where H_{lr} is the cross power of the bandpass noise, H_{ll} and H_{rr} are the auto powers at the two ear entrances, τ_ξ is the maximum interaural delay depending on ξ, and

$$\Delta w_c = 2\pi(f_2 + f_1),$$
$$\Delta w = 2\pi(f_2 - f_1). \tag{3.28}$$

For the calculation of the sound fields with any spectrum signal, the needed data have been reported by Nakajima, Yoshida, and Ando (1993); see also Ando (1985).

The interaural delay time at which the IACC is defined, as shown in Figure 3.7, is the τ_{IACC}. Thus, both the IACC and τ_{IACC} may be obtained from the condition of

$$\frac{\partial \phi_{lr}(\tau)}{\partial \tau} = 0.$$

In the simple sound field described by Equation (3.27), the τ_{IACC} corresponds to the interaural time delay for the horizontal angle ξ defined by τ_ξ. When τ_{IACC} is zero (one of the preferred conditions), then usually a frontal sound image and well-balanced sound field are perceived.

The width of the interaural cross-correlation function defined by the interval of delay time at a value of δ below the IACC, corresponding to the JND of the IACC, is given by the W_{IACC} (Figure 3.7). Thus, the apparent source width (ASW) may be perceived as a directional range corresponding mainly to the W_{IACC}. For the sound field with $\tau_{IACC} = 0$, for example, the term of sin Z/Z in Equation (3.27) is nearly unity, because $Z = \Delta\omega\tau/2$ is small enough, so that

$$W_{IACC} \approx \frac{4}{\Delta\omega_c} \cos^{-1}\left(1 - \frac{\delta}{IACC}\right). \tag{3.29}$$

A well-defined directional impression corresponding to the interaural time delay τ_{IACC} is perceived when listening to sound with a sharp peak in the interaural cross-correlation function with a small value of W_{IACC}. On the other hand, when listening to a sound field with a low value of IACC < 0.15, then subjectively diffuse sound is perceived.

Therefore, these four factors, the geometric mean of sound energies at the two ears, the IACC, τ_{IACC}, and W_{IACC}, are independently related to the space oriented subjective attributes, for example, the subjective diffuseness, the image shift, and the ASW. Further discussion in this area appears in Section 6.1.

3.5. Simulation of Sound Field

According to Equation (3.14), sound fields in a room may be simulated by taking the directional information of its sound source and of its early reflections into consideration (Ando, 1985).

An example of the block diagram of the simulation system for the direct sound and two early reflections and diffused reverberation is shown in Figure 3.8 (Ando et al., 1973), which was used in all subjective judgments experiments and in recording the electro-physiological responses described in this book. In order to realize a small value of IACC, the directions of four loudspeakers for subsequent reverberation (Rev.) are chosen about 55° from the median plane, and the incoherent reverberation signals supplied to the loudspeakers were delayed by Δt_j ($j = 1, 2, 3$).

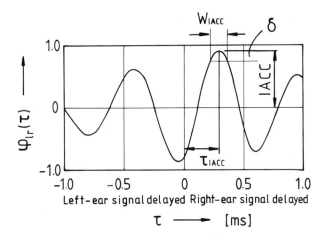

FIGURE 3.7. Definitions of the IACC, τ_{IACC} and W_{IACC} for the interaural cross-correlation function.

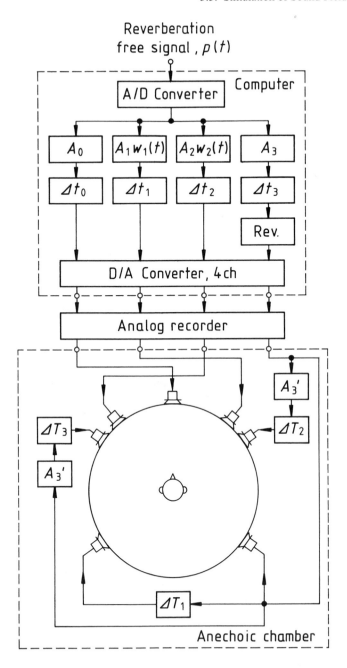

FIGURE 3.8. A simulation system of sound fields.

4

Subjective Preference as an Overall Impression of the Sound Field

A difficulty arises in the investigation of subjective attributes for the sound field in a room, because the sound field consists of a great number of reflections. However, the fundamental attributes are contained in the simplest sound field, which consists of the direct sound, and a single reflection as representative of a set of reflections. The first part of this chapter describes the results of subjective preference studies in relation to the temporal factor, Δt_1, and the spatial factor, IACC, of the sound field. Then, the orthogonal properties of the four acoustic factors are described for the sound field in the room, including the reverberation time.

After obtaining optimum design objectives, the theory of subjective preference is derived. Based on this theory, an example of calculating subjective preference at each seat is demonstrated. In order to examine the theory, subjective preference judgments were conducted in an existing hall without the subjects changing seats in a paired-comparison judgment.

4.1. Subjective Preference of the Simple Sound Field

4.1.1. Preferred Delay Time of a Single Reflection

The sound field consists of the direct sound $\xi_0 = 0°(\eta_0 = 0°)$ and a single reflection from a fixed direction $\xi_1 = 36°(\eta_1 = 9°)$. These angles were selected since they are typical in a concert hall. The delay time Δt_1 was adjusted in the range of 6 ms to 256 ms. Paired comparison tests were performed for all pairs in an anechoic chamber using normal hearing subjects with two different music motifs A and B (Table 3.1). The normalized scores of the sound fields as a function of the delay are shown in Figure 4.1 (Ando, 1977; see also Kang and Ando, 1985).

Obviously, the most preferred delay time, with the maximum score, differs greatly between the two motifs. When the amplitude of reflection $A_1 = 1$, the most preferred delays are around 130 ms and 35 ms for motifs A and B, respectively. It is found that this corresponds to effective durations of the ACF of source signals of 127 ms (motif A) and 35 ms (motif B).

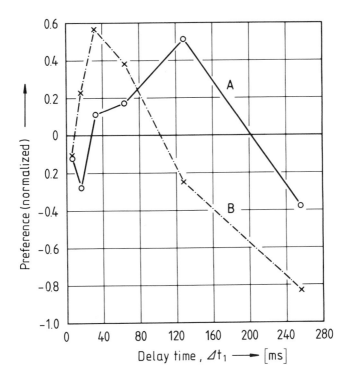

FIGURE 4.1. Preference scores of the sound fields as a function of the delay, $A_1 = 1$ (6 sound fields and 13 subjects (Ando, 1977)). Preference scores are directly obtained for each pair of sound fields by giving $+1$ and -1, corresponding to the positive and negative judgments, respectively. Also, the normalized score is obtained by accumulating the scores for all sound fields (F) tested and all subjects (S), and then dividing by the factor $S(F - 1)$. A: Music Motif A, Royal Pavane by Gibbons, $\tau_e = 127$ ms; and B: Music Motif B, Sinfonietta, Opus 48, III movement, by Malcolm Arnold, $\tau_e = 35$ ms.

After inspection, the preferred delay is found roughly at a certain duration of the ACF, defined by τ_p, such that the envelope of the ACF becomes $0.1A_1$. Thus, $\tau_p = \tau_e$ only when $A_1 = 1.0$. The data collected as a function of the duration τ_p are shown in Figure 4.2, where data from a continuous speech signal of $\tau_e = 12$ ms are also plotted (see also Section 6.2.3). When the envelope of the ACF is exponential, then it is expressed approximately by (Ando, 1985)

$$\tau_p = [\Delta t_1]_p \approx [1 - \log_{10} A_1]\tau_e. \tag{4.1}$$

It is worth noticing that the amplitude of reflection relative to that of the direct sound should be measured by the most accurate method, for example, the square root of the ACF at the origin of the delay time.

Two reasons can be given for explaining why the preference decreases for the short delay range of reflection, $0 < \Delta t_1 < \tau_p$ (Figure 4.3):

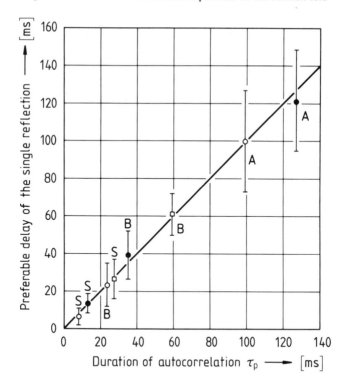

FIGURE 4.2. Relationship between the preferred delay of a single reflection and the duration of the ACF such that $|\phi_p(\tau)|_{\text{envelope}} = 0.1A_1$. Range of the preferred delays are graphically obtained at 0.1 below the maximum score. A, B, and S refer to Music Motif A, Motif B, and speech, respectively. Different symbols indicate the center values obtained at the reflection amplitudes of +6 dB (○), 0 dB (●), and −6 dB (□), respectively (13 to 19 subjects).

(1) tone coloration effects occur because of the interference phenomenon in the coherent time region (Section 6.2.3); and

(2) the IACC increases when Δt_1 is near 0.

On the other hand, echo disturbance effects can be observed when Δt_1 is greater than τ_p.

4.1.2. Preferred Direction of a Single Reflection

The delay time of the reflection, in the experiment showing the preferred direction of a single reflection, was fixed at 32 ms. The direction was specified by loudspeakers located at $\xi_0 = 0°(\eta_0 = 27°)$ and $\xi_1 = 18°, 36°, \ldots, 90°(\eta_1 = 9°)$.

Results of the preference tests are shown in Figure 4.4. No fundamental differences are observed between the curves of the two motifs, in spite of the great

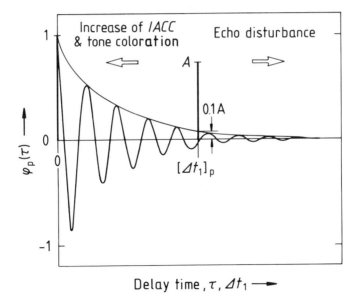

FIGURE 4.3. Subjective attributes before and after the preferred delay time of reflection $[\Delta t_1]_p (= \tau_p)$.

difference of τ_e. The preferred score increases roughly with decreasing IACC. The correlation coefficient between the score and the IACC is -0.8 (at 1% significance level: $p < 0.01$). The score with motif A at $\xi_1 = 90°$ drops to a negative value, indicating that the lateral reflections, coming only from around $\xi_1 = 90°$, are not always preferred. The figure shows that there is a preference for angles less than $\xi_1 = 90°$, and on average there may be an optimum range centered at about $\xi_1 = 55°$. Similar results can be seen in the data from speech signals (Ando and Kageyama, 1977; see also Section 6.4).

4.2. Orthogonal Properties of Acoustic Factors

In order to examine the independence of the effects of the four physical factors on subjective preference judgments and to make continuous the linear scale value of preference for any sound field, two of the four factors were varied simultaneously while the remaining two were held constant. For convenience, sound fields in a concert hall with the same plan as the Symphony Hall in Boston, as shown in Figure 4.5(a) were simulated. The system simulating the sound fields in concert halls is shown in Figure 3.8 of the previous chapter. A computer program provides the time delay of two early reflections ($n = 1, 2$) and the subsequent reverberation ($n > 2$), relative to the direct sound (Figure 4.5(b)). In order to represent the geometrical size of a similar room, the scale of dimension (SD) is introduced

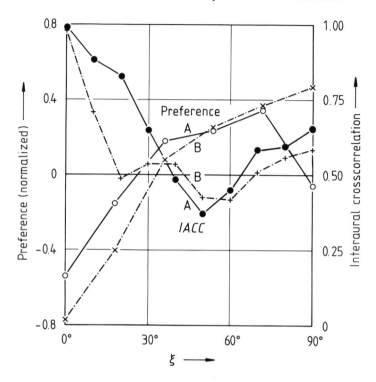

FIGURE 4.4. Preference scores and the magnitude of the IACC of the sound fields as a function of the horizontal angle of a single reflection, $A_1 = 0$ dB (6 sound fields and 13 subjects).

as follows:

$$\Delta t_1 = 22 \text{ (SD)}, \qquad \Delta t_2 = 38 \text{ (SD)}, \qquad \Delta t_3 = 47 \text{ (SD)} \quad \text{[ms]} \qquad (4.2)$$

The reverberation signal with constant frequency characteristics was generated by the Schroeder Reverberator (Schroeder, 1962). To obtain a natural sound, the conditions of the simulation system were carefully selected.

Test A. In order to examine whether or not the temporal-monaural factors influence the scale values of subjective preference independently, paired-comparison tests of the 16 sound fields for each source signal were conducted for changes of SD and T_{sub} with 9 to 14 subjects (Ando, Okura, and Yuasa, 1982). Both of the factors are closely associated with the left hemisphere of the human brain, as is discussed in Sections 5.2 and 5.3.

Test B. In order to determine the independent influence of the spatial factors, LL and IACC, paired-comparison tests of the 12 sound fields for each source signal were performed with 13 to 14 subjects (Ando and Morioka, 1981). Both factors are

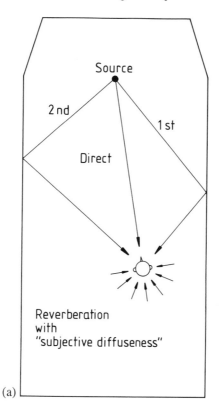

(a)

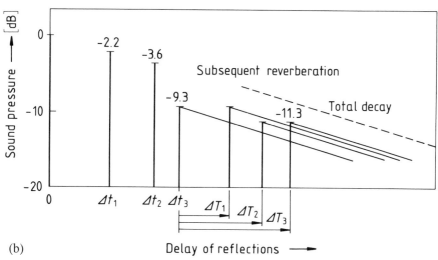

(b)

FIGURE 4.5. (a) A sound field in a concert hall with a plan similar to that of the Symphony Hall, Boston; and (b) amplitude decay of early reflections and subsequent reverberation simulated for subjective preference tests.

closely associated with the right hemisphere of the human brain, as is discussed in Section 5.2.

Test C. For reconfirmation of the independence of the left and right hemispheric factors, T_{sub} and IACC, preference tests were conducted for 16 sound fields with 8 subjects (Ando, Otera, and Hamana, 1983).

Results of the analyses of variance for the scale values obtained by the law of comparative judgments from these three tests are indicated in Table 4.1. According to the significance level, each factor influences the scale value of preference independently (Ando, 1985).

TABLE 4.1.   Analyses of variance for three tests, A, B, and C with 9–16 subjects.

Test	Factor	Sum of squares	Degree of freedom	Mean square	F	Significance level	Contribution [%]
Music Motif A							
	Δt_1 (SD)	0.20	3	0.07	4.4	< 0.05	14
A	T_{sub}	0.73	3	0.24	17	< 0.01	65
	Residual	0.13	9	0.01	—		
	LL	0.99	3	0.33	48	< 0.01	27
B	IACC	2.61	2	1.30	187	< 0.01	71
	Residual	0.04	6	0.01	—		
	T_{sub}	2.44	3	0.82	68	< 0.01	89
C	IACC	0.17	3	0.06	5	< 0.05	5
	Residual	0.11	9	0.01	—		
Music Motif B							
	Δt_1 (SD)	1.20	3	0.40	22	< 0.01	13
A	T_{sub}	7.63	3	2.54	141	< 0.01	84
	Residual	0.16	9	0.02	—		
	LL	0.74	3	0.25	12	< 0.01	24
B	IACC	1.90	2	0.95	47	< 0.01	67
	Residual	0.12	6	0.02	—		
	T_{sub}	2.55	3	0.85	182	< 0.01	79
C	IACC	0.64	3	0.21	46	< 0.01	19
	Residual	0.04	9	0.01	—		

These results hold, even if scale values shifted in origin were applied in analyses.

TABLE 4.2. Examinations on independent effects of each two of four objective factors on the subjective preference judgments.

Factors	LL	Δt_1 (SD)	T_{sub}	IACC
LL	—	Ando and Okada*	None	**Test B**: Ando and Morioka (1981)
Δt_1 (SD)		—	**Test A**: Ando, Okura, and Yuasa (1982)	Ando and Imamura (1979); Ando and Gottlob (1979)
T_{sub}			—	**Test C**: Ando, Otera, and Hamana (1983)

* Unpublished: The effects of Δt_1 were examined under the fixed conditions of a great range of SL (Figure 4.6).

Other Tests. In addition, as listed in Table 4.2, subjective preference judgments were performed in sound fields with multiple early reflections (Ando and Gottlob, 1979), and with subsequent reverberations (Ando and Imamura, 1979). These results confirm that the factors (SD) and the IACC are independent of each other in the subjective preference judgments.

As is discussed in Section 4.3.5, when the sensation level (SL) is weak enough, say 30 dB, the most preferred delay time of the reflection becomes longer than that at the preferred listening level around 80 dB. Thus, it may be concluded that the total scale value of subjective preference is determined by the law of superposition in the range of preferred conditions of the four factors tested. The consistency of the unit of the scale values obtained from the different preference tests has been discussed at length (Ando, 1985).

4.3. Optimum Design Objectives

According to such a systematic investigation of simulating sound fields in a concert hall by the aid of a computer and listening tests (paired-comparison tests), the optimum design objectives and the linear scale value of subjective preference may be described. The optimum design objectives can be described in terms of the subjectively preferred sound qualities, which are related to the temporal and spatial factors describing the sound signals arriving at the two ears. They clearly lead to comprehensive criteria for achieving the optimal design of concert halls as summarized below (Ando, 1983).

4.3.1. Listening Level

The listening level is, of course, the primary criterion for listening to the sound field in concert halls. The preferred listening level depends upon the music and the particular passage being performed. For example, the gross preferred levels obtained by 16 subjects are in the peak ranges of 77 dBA to 79 dBA for Music Motif A (Royal Pavane by Gibbons) with a slow tempo, and 79 dBA to 80 dBA for Music Motif B (Sinfonietta by Arnold) with a fast tempo (see Figure 3.1).

4.3.2. Early Reflections After the Direct Sound

An approximate relationship for the most preferred delay time has been discovered in terms of the autocorrelation function of source signals and the total amplitude of reflections, A (Ando, 1985). Generally, it is expressed by

$$[\Delta t_1]_p = \tau_p, \tag{4.3}$$

$$|\phi_p(\tau)|_{\text{envelope}} \approx kA^c \quad \text{at} \quad \tau = \tau_p, \tag{4.4}$$

where k and c are constants depending on the subjective attributes. (For the important subjective attributes, these constants and the range of values of A tested are listed in Table 6.1.) If the envelope of the ACF is exponential, then

$$\tau_p \approx (\log_{10} \frac{1}{k} - c \log_{10} A)\tau_e, \tag{4.5}$$

where the total pressure amplitude of reflection is given by

$$A = \left[A_1^2 + A_2^2 + A_3^2 + \cdots\right]^{1/2}. \tag{4.6}$$

The relationship of Equation (4.1) for a single reflection may be obtained by putting $A = A_1, k = 0.1$, and $c = 1$ so that

$$\tau_p = [\Delta t_1]_p \approx (1 - \log_{10} A_1)\tau_e.$$

4.3.3. Subsequent Reverberation Time
After the Early Reflections

For flat frequency characteristics of reverberation (one of the preferred conditions), the preferred subsequent reverberation time is expressed approximately by

$$[T_{\text{sub}}]_p \approx 23\tau_e. \tag{4.7}$$

The values A tested were 1.1 and 4.1, that cover the usual conditions of the sound field in a room. A lecture room and conference room must be designed for speech, and an opera house and similar theaters for vocal music. For orchestra music, these may be two or three types of concert-hall designs according to the effective duration of the ACF. For example, Symphony No. 41 by Mozart, "Le Sacre du Printemps" by Stravinsky, and Arnold's Sinfonietta have short ACFs and fit orchestra music of type (A). On the other hand, Symphony No. 4 by Brahms and Symphony No. 7

by Buckner are typical of orchestra music (B). Much longer ACFs are typical for pipe organ music, for example, by Bach.

The most preferred reverberation times estimated for each sound source are shown in Figure 7.7 for the selection of music motifs to be performed. Considering the fact that the value of τ_e is obtained at the ten percentile (or -10 dB) delay of the envelope of the ACF of a source signal, the -60 dB delay time of the ACF-envelope corresponding roughly to the "reverberation time" containing the source signal itself, given by $6\tau_e$. Therefore, the most preferred reverberation time of the sound fields expressed by Equation (4.7) implies about four times the "reverberation time" contained in the source signal itself. Concerning the preferred frequency characteristics, this is discussed in Section 6.2.2.

4.3.4. Dissimilarity of Signals in Both Ears (IACC)

All the available data indicate a negative correlation between the magnitude of the IACC and the subjective preference, i.e., dissimilarity of signals arriving at the two ears is preferred. This holds only under the condition that the maximum value of the interaural cross-correlation function is maintained at the origin of the time delay. If not, then an image shift of the source may occur (Section 4.5). To obtain a small magnitude of IACC in the most effective manner, the directions from which the early reflections arrive at the listener should be kept within a certain range of angles from the median plane, i.e., $\pm(55° \pm 20°)$. It is obvious that the sound arriving from the median plane $\pm0°$ makes the IACC greater. Sound arriving from $\pm90°$ in the horizontal plane is not always advantageous, because the similar "detour" paths around the head to both ears cannot decrease the IACC effectively, particularly for frequency ranges higher than 500 Hz. For example, the most effective angles for the frequency ranges of 1 kHz and 2 kHz are about $\pm55°$ and $\pm36°$, respectively (Figure 6.3).

4.3.5. Effects of Sensation Level on the Preferred Delay Time of Reflections

The purpose of this section is to understand the effects of the sensation level on subjective preference judgments upon a change of temporal factors. As a typical example, the effects on the preferred Δt_1 under different fixed sensation levels (SL) were examined (Ando and Okada, data). The background noise in the listening room was 17.5 dBA. The number of subjects was fifteen. The source signal used here was Music Motif B. But a different part of motif B (with $(\tau_e)_{min} = 60$ ms) from the other experiments described in this book was used, thus the calculated value of the most preferred delay time $[\Delta t_1]_p$ is 60 ms, when $A = 1$.

The results of the scale value as a function of Δt_1 are shown in Figure 4.6 for each fixed value of SL. The preferred values of Δt_1 obtained here as a function of SL are shown in Figure 4.7 normalized to the calculated value $[\Delta t_1]_p = 60$ ms. Obviously, when SL $= 80$ dB, the most preferred value turns out to be the same as

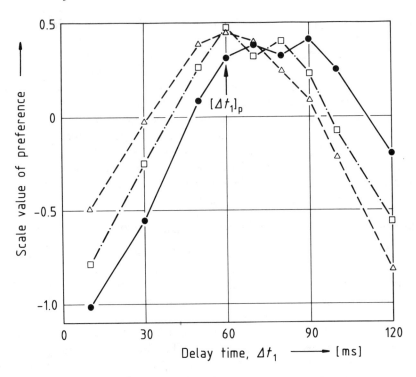

FIGURE 4.6. Scale values as a function of the Δt_1, for fixed values of the sensation level (SL). ($\triangle$): SL = 80 dB; ($\blacksquare$): SL = 55 dB; and ($\bullet$): SL = 30 dB (Ando & Okada, unpublished).

shown in Section 4.3.2. It is found that the most preferred value of Δt_1 increases with decreasing value of SL. Since the SL is considered to be based on the internal biological noise, a similar tendency may result due to the signal to noise ratio in the physical sound field as well. It is quite natural that, if the noise level is increased, Δt_1 and T_{sub} take on larger values, making these temporal factors effective.

4.4. Theory of Calculating Scale Values of Subjective Preference

4.4.1. Theory of Subjective Preference

Let us now put these results into practice. Since the number of orthogonal acoustic factors which are included in the sound signals at both ears are limited, as mentioned in Section 3.4, the scale value of any one-dimensional subjective response

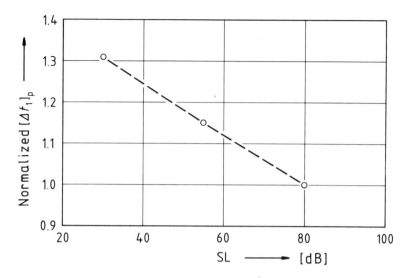

FIGURE 4.7. Normalized values of the preferred delay time as a function of SL, to that at SL = 80 dB.

may be expressed by

$$S = g(x_1, x_2, \ldots, x_I). \tag{4.8}$$

In this study, the linear scale value of preference obtained by the law of comparative judgment is described. It has been verified by a series of experiments that four objective factors act independently of the scale value; when changing two of the four factors simultaneously as indicated in Table 4.2. Results indicate that the units of scale values are almost constant (even if different music motifs are used (Ando, 1985)), so that we may add scale values to obtain the total scale value (Ando, 1983),

$$\begin{aligned} S &= g(x_1) + g(x_2) + g(x_3) + g(x_4) \\ &= S_1 + S_2 + S_3 + S_4, \end{aligned} \tag{4.9}$$

where S_i, $i = 1, 2, 3, 4$, is the scale value obtained relative to each objective parameter. Equation (4.9) indicates a four-dimensional continuity.

The dependence of the scale values on each objective parameter is shown graphically in Figure 4.8. From the nature of the scale value, it is convenient to put a zero value at the most preferred conditions, as shown in Figure 4.8. The results of the scale value of subjective preference obtained from the different test series, using different music programs, yield the following common formula:

$$S_i \approx -\alpha_i |x_i|^{3/2}, \qquad i = 1, 2, 3, 4, \tag{4.10}$$

where the values of α_i are weighting coefficients as listed in Table 4.3. If α_i is

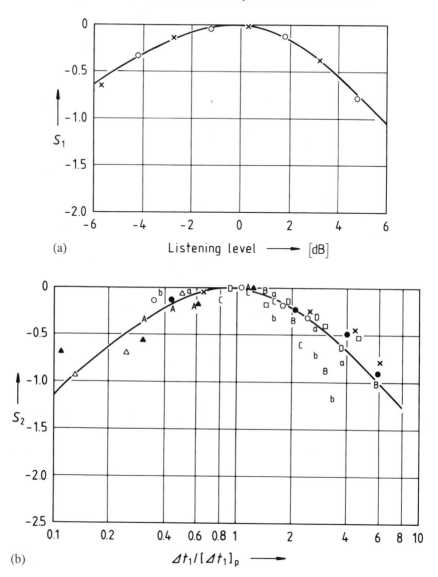

FIGURE 4.8A, B. Scale values of the subjective preference obtained for the simulated sound field in an anechoic chamber. Different symbols indicate scale values obtained by different source signals (Ando, 1985). Even if different signals are used a consistency of scale values as a function of each factor is observed, fitting a single curve. (a) As a function of the listening level, LL. The most preferred listening level, $[LL]_p = 0$ dB; and (b) as a function of $\Delta t_1/[\Delta t_1]_p$.

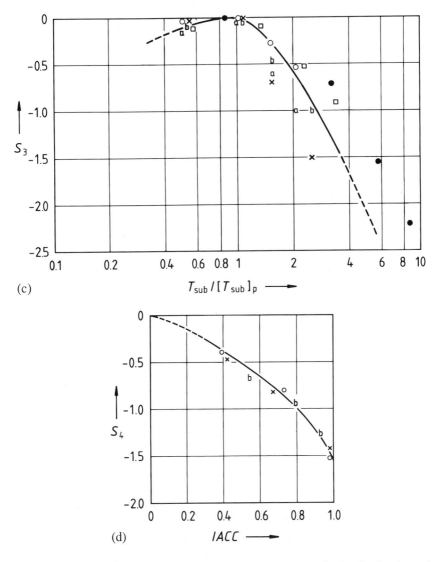

FIGURE 4.8C, D. Scale values of the subjective preference obtained for the simulated sound field in the anechoic chamber. Different symbols indicate scale values obtained by different source signals (Ando, 1985). Even if different signals are used a certain consistency of scale values as a function of each factor is observed, fitting a single curve. (c) as a function of $T_{sub}/[T_{sub}]_p$; and (d) as a function of the IACC. The most preferred values $[\Delta t_1]_p$ and $[T_{sub}]_p$ are calculated by Equations (4.5) with $k = 0.1$ and $c = 1$ and (4.7), respectively.

TABLE 4.3.   Objective parameters and coefficients α_i obtained with 9–16 subjects.

		α_i	
i	x_i	$x_i \geq 0$	$x_i < 0$
1	$20 \log P - 20 \log[P]_p$ (dB)	0.07	0.04
2	$\log(\Delta t_1/[\Delta t_1]_p)$	1.42	1.11
3	$\log(T_{sub}/[T_{sub}]_p)$	$0.45 + 0.75A$	$2.36 - 0.42A$
4	IACC	1.45	—

close to zero, then a lesser contribution of the factor x_i on subjective preference is signified.

The factor x_1 is given by the sound pressure level difference, measured by the A-weighted network, so that

$$x_1 = 20 \log P - 20 \log[P]_p, \tag{4.11}$$

P and $[P]_p$ being the sound pressure at a specific seat, and the most preferred sound pressure that may be assumed at a particular seat position in the room under investigation;

$$x_2 = \log\left(\frac{\Delta t_1}{[\Delta t_1]_p}\right), \tag{4.12}$$

$$x_3 = \log\left(\frac{T_{sub}}{[T_{sub}]_p}\right), \tag{4.13}$$

$$x_4 = \text{IACC}. \tag{4.14}$$

Thus, the scale values of preference have been formulated approximately in terms of the 3/2 power of the normalized objective parameters, expressed in the logarithm for the parameters, x_1, x_2, and x_3. The spatial binaural parameter x_4 is expressed in terms of the 3/2 power of its real values, indicating a greater contribution than those of the temporal parameters. Thus, the scales values are not greatly changed in the neighborhood of the most preferred conditions, but decrease rapidly outside of the ranges. Since the experiments were conducted to find the optimal conditions, this theory holds only in the ranges of the preferred conditions tested for the four factors.

4.4.2. Calculation of the Subjective Preference at Each Seat

As a typical example, we will discuss the quality of the sound field at each seat in a concert hall with a shape similar to that of the Symphony Hall in

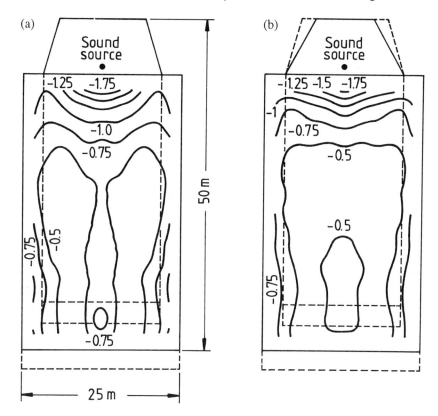

FIGURE 4.9. An example of calculating scale values with the four factors using Equations (4.9)–(4.14). (a) Contour lines of the total scale value for Boston Symphony Hall with original side reflectors on the stage; and (b) contour lines of the total scale values for the side reflectors optimized.

Boston. Suppose that a single source is located at the center, 1.2 m above the stage floor. Receiving points at a height of 1.1 m above the floor level correspond to the ear positions. Reflections with their amplitudes, delay times, and directions of arrival at the listeners are taken into account using the image method.

Contour lines of the total scale value of preference calculated for Music Motif B are shown in Figure 4.9. This figure demonstrates the effects of the reflection from the side reflections on the stage. The side wall on the stage may produce decreasing values of IACC for the audience area. Thus, the preference value at each seat is increased, as is shown in Figure 4.9(b) in comparison with that in Figure 4.9(a).

In this calculation, the reverberation time is assumed to be 1.8 s throughout the hall and the most preferred listening level, $[LL]_p = 20 \log[P]_p$ in Equation (4.11), is set for a point on the center line 20 m from the source position.

4.5. Examination of Subjective Preference in an Existing Hall

4.5.1. Preference Test in an Existing Hall

The subjective preference judgments for different source locations on the stage were performed by the paired-comparison tests at each set of the seats. The relationship between the resulting scale value of subjective preference and the physical factors obtained by simulation, using an architectural plan drawings, was examined by factor analysis (Sato, Mori, and Ando, 1997). Calculated scale values of subjective preference were reconfirmed in an existing concert hall, the Uhara Hall in Kobe (Figure 4.10). The physical factors at each set of seats for the four source locations on the stage (Figure 4.10(a)) were calculated. In the simulation, the directional characteristics of four loudspeakers used in preference tests were taken into consideration. The simulation calculation was performed up to three reflection times for each directional reflection n in Equation (3.13) to a listener. Due to a floor structure with a fair amount of acoustic transparency, the floor reflection was not taken into account for the calculation, and part of the diffuser ceiling was regarded as a nonreflective plane for the sake of convenience. In the calculation of the IACC, the listeners faced toward the center of the stage, so that the IACC was not always a maximum at the interaural time delay $\tau = 0$.

This hall contains 650 seats with a volume of 4870 m^3. Loudspeakers were placed at 0.8 m above the stage floor, and 64 listeners divided into 21 groups were seated at the specified set of seats. Without moving from seat to seat and excluding the effects of other physical factors such as visual and tactual senses on judgments, subjective preference tests by the paired-comparison method were conducted, switching only the loudspeakers on the stage. As a source signal, Music Motif B was selected in the tests. Scale values of preference were obtained by applying the law of comparative judgment (Case V; Thurstone, 1927; Torgerson, 1958) and were reconfirmed by the goodness of fit (Mosteller, 1951). In order to obtain enough data, a set of adjoining three or four seats was chosen in a single test session. This session was repeated five times, exchanging the sets of seats, and 14 to 16 subjects in total were tested at each set of seats.

4.5.2. Results of Multiple-Dimensional Analysis

In order to examine the relationship between the scale values of subjective preference and physical factors obtained by simulation of an architectural scheme, the data were analyzed by the factor analysis described in Appendix I (Hayashi, 1952; 1954a and b).

Of the four physical factors, the reverberation time was almost constant for the source location and the seat location throughout the hall, and thus not involved in the analysis. As was previously discussed, as a condition of calculating the scale value of preference, and the maximum value of the interaural cross-correlation

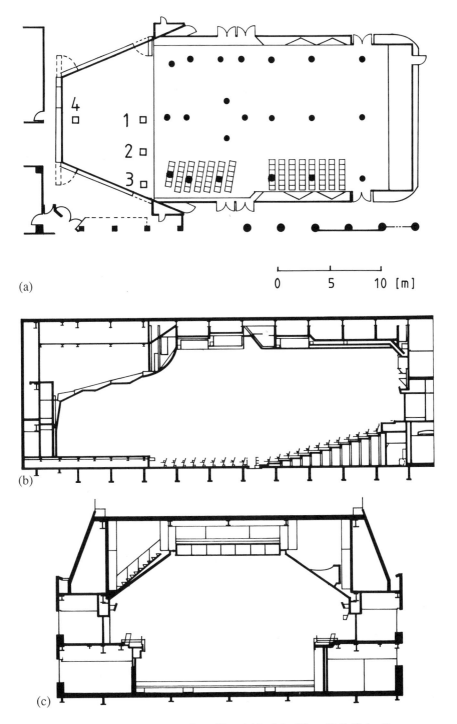

(a)

0   5   10 [m]

(b)

(c)

FIGURE 4.10. Plan (a) and cross-sections (b) and (c) of the Uhara Hall, Kobe. Four source positions on the stage, which were switched in the paired-comparison tests of preference without moving subjects from seat to seat. Listening positions were 21 locations and included neighboring seats. Illustration of seats is shown only part of Figure 4.10(a).

function must be maintained at an interaural time delay $\tau = 0$ to ensure frontal localization of the sound source. However, the IACC was not always maintained at $\tau = 0$ due to the loudspeaker locations, because the subjects could not always be facing the source location. In this analysis, therefore, the effect of the interaural time delay of the IACC, i.e., τ_{IACC} was added as an additional factor. Thus, the outside variable to be predicted with factor analysis was the scale value obtained by subjective judgments, and the explanatory factors were:

(1) the listening level;
(2) the initial time-delay gap;
(3) the IACC; and
(4) the interaural time delay of the IACC (τ_{IACC}).

The scores for each category of factors obtained from the factor analysis are shown in Figure 4.11. As shown in Figure 4.11(a), the scores of the listening level indicate a peak at the subcategory of 83 dB to 85.9 dB and decrease the score apart from the preferred listening level. For the IACC, the preference score increases with a decrease in the IACC (Figure 4.11(c)). It is worth noticing that the scores of the above-mentioned two factors are in good agreement with the preference scale values obtained by preference judgments for a simulated sound field. The scores of the initial time delay gap which are normalized to the optimum value ($\Delta t_1/[\Delta t_1]_p$) peaked at a smaller value than the most preferred value of the initial time-delay gap obtained from the simulated sound fields (Figure 4.11(b)). It is considered that, due to the limited range of the Δt_1 in the existing concert hall and the limited data in the short range of the Δt_1, the effects of the Δt_1 of the sound fields were rather minor in this investigation. Concerning τ_{IACC} (Figure 4.11(d)), the score decreases monotonically as the delay is increased. This may be caused by an image shift.

The relationship between the scale value obtained by subjective judgments and the total score at each center of three or four seats is shown in Figure 4.12. The scale values of preference are well predicted with the total score for four loudspeaker locations ($r = 0.70$, $p < 0.01$). On occasion there is a certain degree of coherence between physical factors, for example, the calculated listening level and the IACC for sound fields in existing concert halls. However, due to the fact that the factors are theoretically orthogonal, the preference scores obtained here are in good agreement with the calculated preference scales that are obtained by simulating sound fields. It is possible that such an apparent coherence may be eliminated without loss of any information, even if some data are excluded for the analyses.

In this study, the subjective preference of source locations on the stage are examined at each set of seats. The rear source (#4) is more preferred than that of the other sources. The side source (#3) indicates low preference, due to the interaural time delay of the IACC. The initial time-delay gap has a small influence on the total score.

This study introduces the effects of the interaural time delay of the IACC on the preference. If the IACC is obtained at a certain interaural time delay, then scores decrease rapidly. The results of the analyses described here demonstrate that the

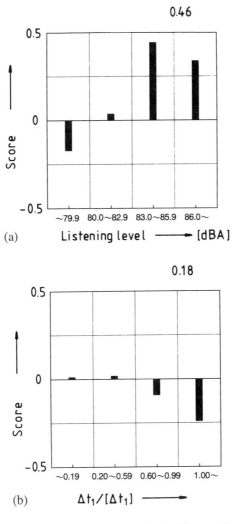

FIGURE 4.11A, B. Scores for each category of four physical factors obtained by factor analysis. The number signifies the partial correlation coefficient between the score and each factor. (a) Listening level; (b) normalized initial time-delay gap between the direct sound and the first reflection.

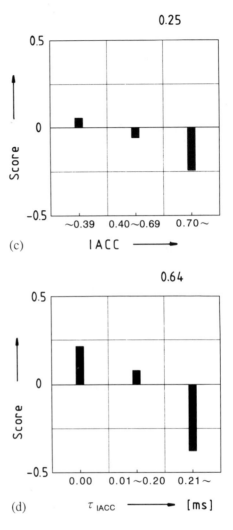

FIGURE 4.11c, d. Scores for each category of four physical factors obtained by factor analysis. The number signifies the partial correlation coefficient between the score and each factor. (c) IACC; and (d) interaural time delay of the IACC, τ_{IACC}, found as the most significant factor in this investigation with loudspeaker production on the stage. Tendencies obtained here are similar to those of the scale values shown in Figure 4.8, Section 4.4.

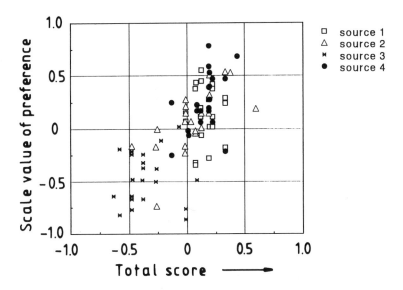

FIGURE 4.12. Relationship between the scale values obtained by paired-comparison tests in the existing hall and the total scores calculated by factor analysis using the scores shown in Figure 4.11. The correlation coefficient, $r = 0.70 (p < 0.001)$.

theory of calculating subjective preference by the use of four physical parameters is supported only when the maximum value of interaural cross-correlation is maintained at $\tau = 0$. This condition is usually obtained in a real concert facing the visible performer.

5

Human Hearing System

The first part of this chapter describes the sensitivity of the human ear to a sound source that is formed primarily by the physical system consisting of the external canal, eardrum, and bone chain with oval window.

Next, in order to describe a background of subjective responses, the electrical-physiological responses of the auditory pathways and of the left- and right-cerebral hemispheres are analyzed. Several remarkable findings are offered in this chapter.

5.1. Physical Systems of Human Ears

5.1.1. Head, Pinna, and External Auditory Canal

The acoustic environment is perceived by the ears, in which a sound signal is given by a time sequence. The three-dimensional space is also perceived by the ears, mainly because the head-related transfer functions $H_{l,r}(r|r_0, \omega)$ between a source point and the two ear entrances have directional qualities from the shapes of the head and the pinna system. The directional information is contained in such head-related transfer functions, including the interaural time difference.

Figure 5.1 shows examples of the amplitude of the head-related transfer function $H(\xi, \eta, \omega)$ as parameters of the angle of incidence ξ ($\eta = 0$). These were measured by the single-pulse method at the far-field condition (Mehrgardt and Mellert, 1977). The angle $\xi = 0°$ corresponds to the frontal direction and $\xi = 90°$ corresponds to the lateral direction toward the side of the ear being examined.

Since the diameter of the external canal is small enough compared with the wavelength (below 8 kHz), the transfer function of $E(\omega)$ is independent of the directions in which sound is incident on the human head for the audio-frequency range:

$$E_{l,r}(\xi, \eta, \omega) \approx E_{l,r}(\omega) \approx E(\omega).$$

Therefore, interaction between the sound field in the external canal and that of the outside, including the pinna, is insignificant. The transfer function from the free

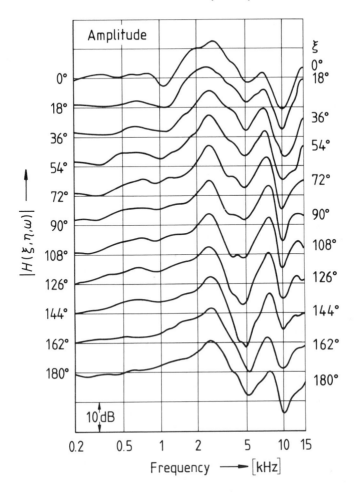

FIGURE 5.1. Transfer functions (amplitude) from a free field to the ear-canal entrance as a parameter of the horizontal angle ξ (Mehrgardt and Mellert, 1977).

field to the eardrum can be obtained by multiplying together the following two functions:

(1) the sound field from the sound source in the free field to the ear-canal entrances, $H_{l,r}(\xi, \eta, \omega)$; and
(2) the sound field from the entrance to the eardrum, $E(\omega)$.

Measured absolute values of $E(\omega)$ are shown in Figure 5.2, where the variations in the curves obtained by different investigations are caused mainly by the different definitions of the ear-canal entrance point. A typical example of the transfer function from a sound source in front of the listener to the eardrum is shown in Figure 5.3. This corresponds to direct sound when the listener is facing the performer. The transfer functions obtained by these three reports (Wiener and Ross, 1946; Shaw, 1975; Mehrgardt and Mellert, 1977) are not significantly different for frequencies up to 10 kHz.

5.1.2. Eardrum and Bone Chain

Behind the eardrum are the tympanic cavities containing the three auditory ossicles, the malleus, incus, and stapes. This area is called the middle ear (Figure 5.4).

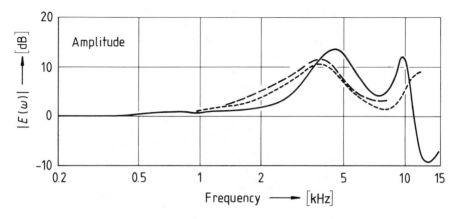

FIGURE 5.2. Transfer functions of the ear canal. (———––—) : from Wiener and Ross (1946); (- - - - - -) : from Shaw (1974); and (———————): from Mehrgardt and Mellert (1977).

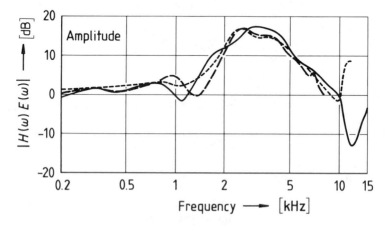

FIGURE 5.3. Transfer functions from a sound source in front of the listener to the eardrum. (———––—) : From Wiener and Ross (1946); (- - - - - -) : from Shaw (1975); (———————): from Mehrgardt and Mellert (1977).

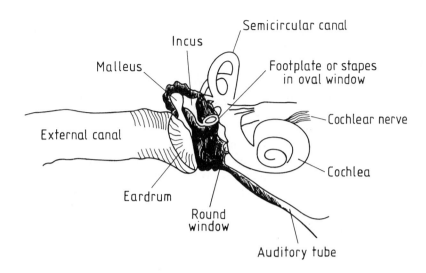

FIGURE 5.4. Schematic illustration of the human ear (modified from Dorland, 1947).

The sound pressure striking the eardrum is transduced into vibration. The middle ear ossicles transmit the vibration to the cochlea. The vibration pattern of the human eardrum was first measured by Békésy (1960) by making a point-by-point examination with an electric capacitive probe. Later, Tonndorf and Khanna (1972) measured the vibration pattern by time-averaged holography, which allows finer vibration patterns on the eardrum to be perceived, as shown in Figure 5.5. Note that the outline of the malleus is visible in the pattern at the value of 3.5. The vibration on the malleus is transmitted to the incus and the stapes. The transfer

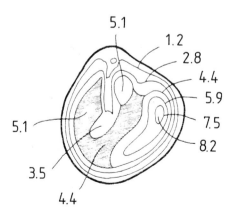

FIGURE 5.5. Contour lines of equal amplitude of human eardrum vibration at 525 Hz (121 dB SPL). Each value should be multiplied by 10^{-5} cm (Tonndorf and Khanna, 1972).

function $C(\omega)$ of the human middle ear between the sound pressure at the eardrum and the apparent sound pressure on the cochlea is plotted in Figure 5.6. The values have been rearranged by the author. Data were obtained by Onchi (1961) and Rubinstein et al. (1966) from cadavers. The maxima at 1 kHz are adjusted to the same value. Later, Puria, Rosowski, and Peake (1993) made measurements by a system that included a hydropressure transducer used in the vestibule as shown in Figure 5.7. The hydropressure transducer and the microphone with identical sound pressure stimuli in air produced estimates of pressure within 0.5 dB for the range of 50 Hz to 11 kHz. The results at the sound pressure levels of 106 dB, 112 dB, and 118 dB indicating similar values are shown in Figure 5.8. These results agree well with the data shown in Figure 5.6, as far as relative behavior is concerned. The transfer function measured at 124 dB showed some signs of nonlinearity, but below about 118 dB it was consistent with a linear system. The magnitude of the middle-ear pressure gain is about 20 dB in the frequency range 500 Hz to 2 kHz.

For the usual sound field, the transfer function between a sound source located in front of the listener and the cochlea may be represented by

$$S(\omega) = H(0, 0, \omega)E(\omega)C(\omega). \tag{5.1}$$

The values are plotted in Figure 5.9 with data from Onchi (1961) and Rubinstein et al. (1966). The pattern of the transfer function agrees with the ear sensitivity for people with normal hearing ability, so that the ear sensitivity can be characterized primarily by the transfer function from the free field to the cochlea (Zwislocki, 1976). Better agreement can be obtained with the values reexamined in the low-frequency range (Berger, 1981).

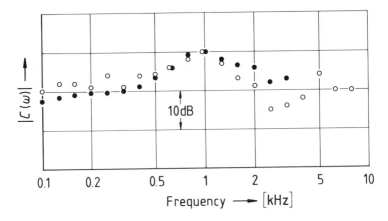

FIGURE 5.6. Transfer function (relative amplitude) of the human middle ear between the sound pressure at the eardrum and the apparent pressure on the cochlea. (●): Average value measured (modified from Onchi, 1961); (○): Measured value (Rubinstein et al., 1966).

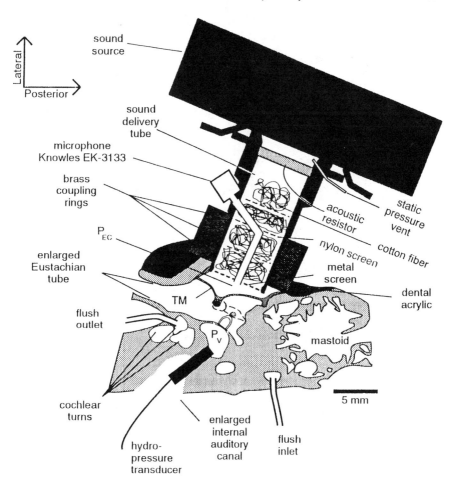

FIGURE 5.7. Measurement system of middle-ear transfer function (Puria, Rosowski, and Peake, 1993). To measure the inner-ear pressure, a hydropressure transducer was placed in the vestibule facing the stapes. In order to ensure that the cochlea remains fluid filled during the measurement, an inlet flush tube was cemented into the upper semicircular canal and an outlet flush tube was cemented into the apical turn of the cochlea.

5.1.3. The Cochlea

The stapes is the last bone of the three auditory ossicles, and is the smallest bone of the human body. It is connected with the oval window, and drives the fluid in the cochlea, producing a traveling wave along the basilar membrane. The cochlea contains the sensory receptor organ on the basilar membrane, which transforms the fluid vibration into the neural code, Figure 5.10. The basilar membrane is so flexible that each section can move independently of the neighboring section.

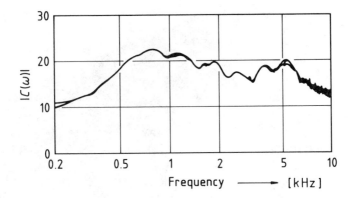

FIGURE 5.8. Transfer function of the human middle ear between the sound pressure at the eardrum and the inner-ear pressure (Puria, Rosowski, and Peake, 1993). The global behavior is surprisingly similar to that of Figure 5.6.

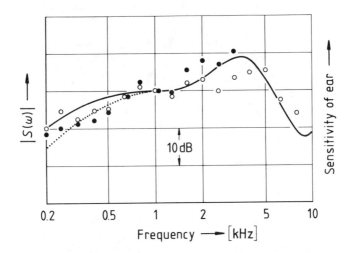

FIGURE 5.9. Sensitivity of the human ear to a sound source in front of the listeners. (————): Normal hearing threshold (ISO recommendation); (- - - - - -): reexamined in the low-frequency range (Berger, 1981); (●,○): transformation characteristics between the sound source and the cochlea, $S(\omega) = H(\omega)E(\omega)C(\omega)$; (●): data obtained from measured values $C(\omega)$ by Onchi (1961); and (○): from Rubinstein (1966), which are combined with the transfer function $H(\omega)E(\omega)$ as measured by Mehrgardt and Mellert (1977).

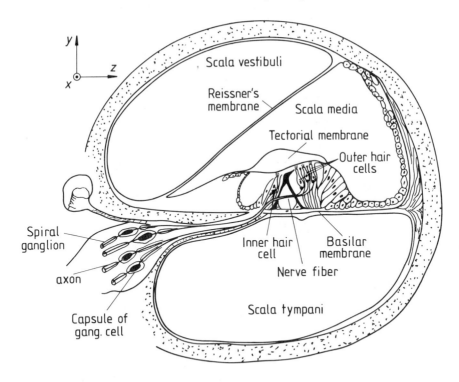

FIGURE 5.10. Cross-section through the cochlea showing the fluid filled canals and the basilar membrane supporting hair cells (modified from Rasmussen, 1943).

The traveling waves on the basilar membrane observed by Békésy (1960), Figure 5.11(a, b), are consistent with this representation.

5.1.4. The Nervous System

The nervous system is one of the most important parts in the whole acoustic system. Obviously, there are some deep connections between sound signals, responses of the auditory system and brain, and subjective attributes, and will be covered in Sections 5.2 and 5.3.

The mechanical information in the traveling waves on the basilar membrane is transduced into biological information. The transducers, consisting of about 15,000 receptors on the basilar membrane, are specialized nerve cells called hair cells. The action potentials from the hair cells are conducted and transmitted to a higher level in the brain. The frequency response curve, called the "tuning curve" of a single fiber, were first systematically demonstrated in the auditory pathway by Katsuki and his group (1958). The results of the threshold response in the potential activity of the cochlear nerve of a cat are shown in Figure 5.12(a), and of the trapezoid body in Figure 5.12(b). The important phenomenon is the so-called

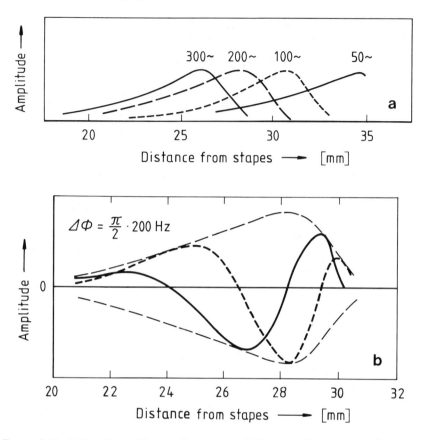

FIGURE 5.11. (a) Envelope of the traveling wave; and (b) the traveling waves on the basilar membrane at 200 Hz (Békésy, 1960).

sharpening effect. The tuning curve becomes sharper than the resonance curve on the basilar membrane. This tendency becomes more distinct at higher levels and the slope reaches the order of 1000 dB/octave. Békésy (1967) explained this as a result of a lateral inhibition action of neural networks. Interactions between neighboring neurons are responsible, at least partially, for the sharpening. Therefore, responses of a single pure tone ω tend to approach a limited region in the auditory pathway x'. Accordingly, the input power density spectrum of the cochlea $I(\omega)$ can be roughly mapped at the nerve position x', so that the spectrum can be written as $I(x')$. This neural activity appears capable of attaining the autocorrelation function as described in Section 3.1.

In addition to the cochlea nuclei, there are the superior olivary complex, the lateral lemniscus nuclei, the inferior colliculus, and the medial geniculate body. Neural signals are processed at every relay station. Since several interaural cross connections are known to exist as physiological structures (e.g., Pickles, 1982), it

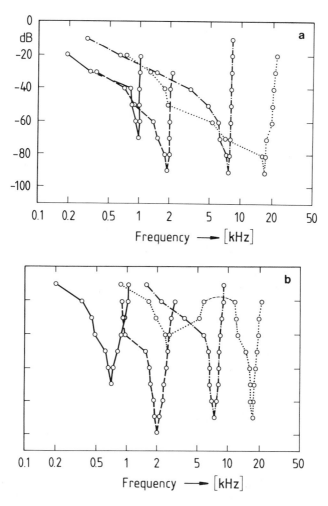

FIGURE 5.12. Frequency response functions of single fibers as threshold responses in the potential activity of a cat's auditory system. Each line indicates the response of different single fibers (Katsuki, Sumi, Uchiyama, and Watanabe, 1958). (a) Cochlea nerve; and (b) trapezoid body.

is quite possible that there exists an interaural cross-correlation mechanism at the inferior colliculus as discussed below. Also, in the following section, the results of some experiments with records of electro-physiological responses from the auditory pathways and the left and right cerebral hemispheres will be described.

5.2. Influence of Electro-Physiological Responses from Auditory Pathways and Human Cerebral Hemispheres Relating to Subjective Preference

5.2.1. Auditory Brainstem Responses (ABR)

A possible mechanism has been assumed for the IACC in the auditory pathways in judging subjective preference and subjective diffuseness. The left and right auditory–brainstem responses (ABR) were recorded in order to justify such a mechanism for the spatial information that might exist in the auditory pathways (Ando, Yamamoto, Nagamatsu, and Kang, 1991).

(A) ABR Recording and Flow of Neural Signals

As a source signal $p(t)$, a short-pulse signal (50 μs) was supplied to a loudspeaker with frequency characteristics of ± 3 dB for 100 Hz to 10 kHz. This signal was repeated every 100 ms for 200 s (2000 times), and left and right ABRs were recorded through electrodes placed on the vertex, and left and right mastoids. The distance $|r - r_0|$ between the loudspeakers and the center of the head was kept at 68cm $\pm$ 1 cm. The loudspeakers were located on the right-hand side of the subject.

Examples of recording ABR as a parameter of the horizontal angle of sound incidence of one of four subjects are shown in Figure 5.13. It is seen that waves I–VI from the vertex and the right mastoid differ in amplitude as indicated by each curve. Quite similar ABR data for the four subjects who participated were obtained, and data for the four subjects (23 $\pm$ 2 years of age, male) were averaged. As shown in Figure 5.14(a) (wave I), of particular interest is the fact that amplitudes from the right which may correspond to the sound pressure from the source located at the right-hand side are greater than those from the left, $r > l$ for $\xi = 30° - 150°$, ($p <$ 0.01). This tendency is reversed in wave II as shown in Figure 5.14(b), $l > r$ for $\xi = 60°$ and 90°, $p < 0.05$. The behavior of wave III shown in Figure 5.14(c) is similar to that in wave I, $r > l$ for $\xi = 30° - 150°$, $p < 0.01$. This tendency is again reversed in wave IV as shown in Figure 5.14(d), $l > r$ for $\xi = 60°$ and 90°, $p < 0.05$, and this is maintained further in wave VI as shown in Figure 5.14(f) even though absolute values are amplified, $l > r$ for $\xi = 60°$ and 90°, $p < 0.05$. From this evidence, it is likely that the flow of neural signals is interchanged three times between the cochlear nucleus, the superior olivary complex, and the lateral lemniscus, as shown in Figure 5.15 for this spatial information process. The interchanges at the inferior colliculus may be operative for the interaural signal processing as discussed below.

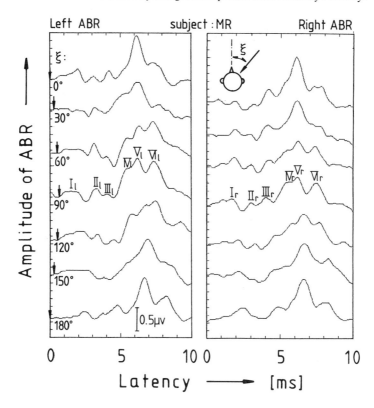

FIGURE 5.13. Examples of the auditory brainstem response (ABR) obtained between the vertex and left- and right-mastoids, as a parameter of the horizontal angle of sound incidence. The abscissa is the time relative to the time when the single pulse arrives at the right ear entrance. Arrows indicate the time delay, depending upon the sound source location of the right-hand side of the subject, and the null amplitude of the ABR.

In wave V, as shown in Figure 5.14(e), such a reversal cannot be seen, and the relative behavior of amplitudes of the left and the right are parallel and similar. Thus, these two amplitudes were averaged and plotted in Figure 5.18 (symbols V). In this figure, the amplitudes of wave IV (left and right, symbols l and r) are also plotted, in reference to the ABR amplitudes at frontal sound incidence.

Concerning latencies of waves I through VI relative to the time when the short pulse was supplied to the loudspeaker, the behaviors indicating relatively short latencies in the range around $\xi = 90°$ were similar (Figure 5.16). It is remarkable that a significant difference is achieved ($p < 0.01$) between averaged latencies at $\xi = 90°$, and those at $\xi = 0°$ (or $\xi = 180°$), i.e., a difference of about 640 μs on average, which corresponds to the interaural time difference of sound incident at $\xi = 90°$. It is most likely that the relative latency at wave III may be reflected by

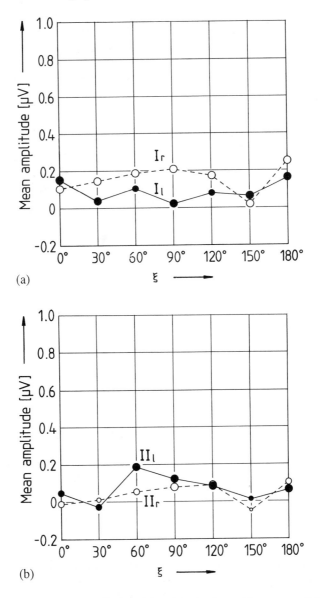

(a)

(b)

FIGURE 5.14A, B. Averaged amplitudes of the ABRs for four subjects, waves I–VI. The size of circles indicated the number of available data from four subjects. (●): Left ABRs; (○): Right ABRs. (a) Wave I and (b) Wave II.

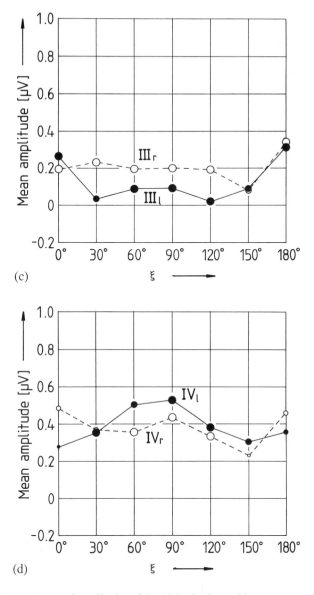

FIGURE 5.14c, d. Averaged amplitudes of the ABRs for four subjects, waves I–VI. The size of circles indicated the number of available data from four subjects. (●): Left ABRs; (○): Right ABRs. (c) Wave III and (d) Wave IV.

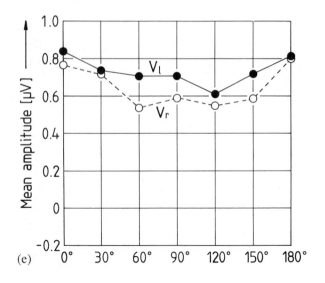

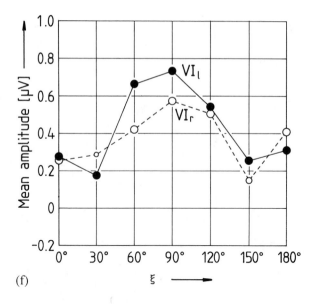

FIGURE 5.14E, F. Averaged amplitudes of the ABRs for four subjects, waves I–VI. The size of circles indicated the number of available data from four subjects. (●): Left ABRs; (○): Right ABRs. (e) Wave V and (f) Wave VI.

ABR WAVES

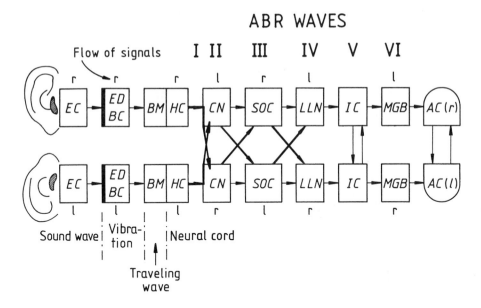

FIGURE 5.15. Schematic illustration of the flow of signals in auditory pathways. EC: external canal; ED and BC: eardrum and bone chain; BM and HC: basilar membrane and hair cell; CN: cochlear nucleus; SOC: superior olivary complex; LLN: lateral lemniscus nucleus; IC: inferior colliculus; MGB: medial denticulate body; AC: auditory cortex of the right and left hemispheres. The source location of each wave (ABR) was previously investigated for both animal and human subjects (Jewett, 1970; Lev and Sohmer, 1972; Buchwald and Huang, 1975).

the interaural time difference. No significant differences could be seen between the latencies of the left and right of waves I–IV, as indicated in Figure 5.16.

(B) ABR Amplitudes in Relation to the IACC

Figure 5.17 shows values of the magnitude of interaural cross-correlation and the autocorrelation functions at the time origin. These were measured at the two ear entrances of a dummy head as a function of the horizontal angle after passing through the A-weighting networks. The averaged amplitudes of wave IV (left and right) and averaged amplitudes of wave V that were both normalized to the amplitudes at the frontal incidence ($\xi = 0°$) are shown in Figure 5.18. Even with a lack of data at $\xi = 0°$, similar results could be obtained when the amplitudes were normalized to those at $\xi = 180°$. Although we cannot make a direct comparison between the results in Figures 5.17 and 5.18, it is interesting to point out that the relative behavior of wave IV(l) in Figure 5.18 is similar to $\Phi_{rr}(0)$ in Figure 5.17 which was measured at the right-ear entrance r. Also, the relative behavior of wave IV$_r$ is similar to $\Phi_{ll}(0)$ at the left-ear entrance l. In fact, the amplitudes of wave IV (left and right) are proportional to $\Phi_{xx}(0)$ ($x = r, l$, respectively), due

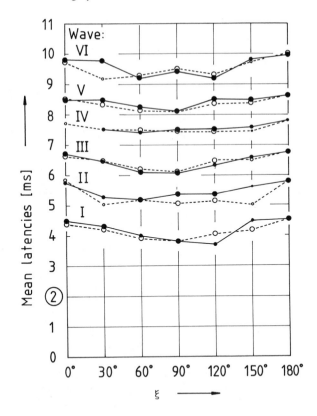

FIGURE 5.16. Averaged latencies of the ABRs for four subjects, waves I–VI. Sizes of the circle indicate the number of available data from four subjects. The latency of 2 ms indicated by two corresponds to the distance between the loudspeaker and the center of the head. (●): left ABRs; (○): right ABRs.

to the interchange of signal flow. The behavior of wave V is similar to that of the maximum value, $|\Phi_{lr}(\tau)|_{max}$, $|\tau| < 1$ ms. Since correlations have the dimensions of the power of the sound signals, i.e., the order of A^2, the IACC defined by Equation (3.23) may correspond to

$$P = \frac{A_V^2}{\left[A_{IV,r} A_{IV,l}\right]},\qquad(5.2)$$

where A_V is the amplitude of the wave V, which may be reflected by the "maximum" neural activity ($A_V^2 \approx |\Phi_{lr}(\tau)|_{max}$) at the inferior colliculus (see Figure 5.15). Also, $A_{IV,r}$ and $A_{IV,l}$ are amplitudes of wave IV on the right and left, respectively. The results obtained by Equation (5.2) are plotted in Figure 5.19. It is clear that the relative behaviors of the IACC and P are in good agreement, except for the value of P at $\xi = 150°$ at which only a single datum for $A_{IV,r}$ was obtained

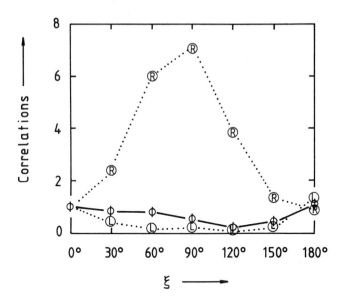

FIGURE 5.17. Correlations of sound signals at the left- and right-ear entrances of a dummy head. L: $\Phi_{ll}(0)$ measured at the left ear; R: $\Phi_{rr}(0)$ measured at the right ear; and Φ: Maximum interaural cross-correlation, $|\Phi_{lr}(\tau)|_{max}$, $|\tau| \leq 1$ ms.

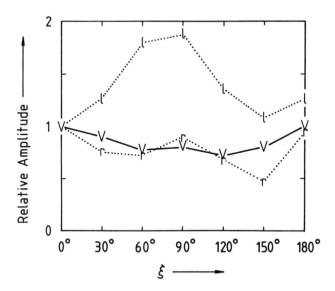

FIGURE 5.18. Averaged amplitudes of waves IV_l (symbol: l) and IV_r (symbol: r), and averaged amplitudes of waves V_l and V_r (symbol: V) normalized to the amplitudes at the frontal incidence (four subjects).

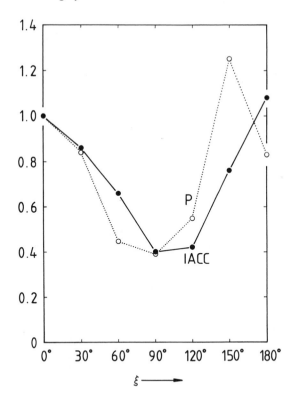

FIGURE 5.19. Values of the IACC and P calculated by Equations (3.23) and (5.2), respectively. A linear relationship between the IACC and the value P is obtained ($p < 0.005$). Note that available datum at $\xi = 150°$ was a single subject.

with only a single subject. The values exceeding unity are caused by the error in the measurements. Obviously, a high correlation between the values of the IACC and P is achieved, i.e., 0.92 ($p < 0.005$).

(C) Remarks

The amplitudes of the ABR clearly differ according to the horizontal angle of the incidence of sound to the listener, as shown in Figure 5.14. In particular, it is found that the amplitudes of waves IV_l and IV_r are nearly proportional to the sound pressures at the right- and left-ear entrances, respectively, when the amplitude is normalized to that at $\xi = 0°$ or $180°$.

As far as the left- and right-amplitude behaviors of the ABR recorded here are concerned, the first interchange of the neural signal is considered to occur at the entrances of the cochlear nucleus, the second interchange may take place at the superior olivary complex, and the third may be at the lateral lemniscus nucleus, as shown in Figure 5.15. Thompson and Thompson (1988), who used neuroanatomical tract-tracing methods in guinea pigs, found four separate pathways connecting one cochlea either with the other cochlea or with itself all via brainstem neurons. This may relate to the first interchange at the entrance of the cochlear nucleus.

As has been discussed the "maximum," of the neural activity for wave V (inferior colliculus) in the auditory pathways corresponds to the IACC appearing in the amplitude of the ABR around 8.5 ms after the sound signal supplied to the loudspeakers (68 cm $\pm$ 1 cm).

The latency of wave V decreases with increasing sensation level as shown in Figure 5.20 (Hecox and Galambos, 1974). This implies binaural summation for the sound energy or the sound pressure level which may be reflected in both $\Phi_{ll}(0)$ and $\Phi_{rr}(0)$ corresponding to Equation (3.24). As will be discussed below, the sound level response indicates right hemisphere dominance. Also, the relative latency at wave III corresponds to the interaural time difference (Figure 5.16).

As described below, it is remarkable that there is a linear relationship between the IACC and the N_2-latency observed in the slow vertex response (SVR) over both cerebral hemispheres ($p < 0.025$) as shown in Figure 5.21 (see also Figure 5.26, right). Thus, the subjective preference and subjective diffuseness judgments of the sound field, described previously in relation to the IACC, are well based on the activities of the auditory–brain system.

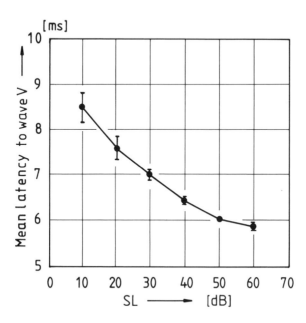

FIGURE 5.20. Latency to Wave V as a function of the sensation level (Hecox and Galambos, 1974). This may correspond to a binaural summation of sound energies from the left and right ears. (See also the P_1 and N_1 latencies of the SVR as shown in Figure 5.26.)

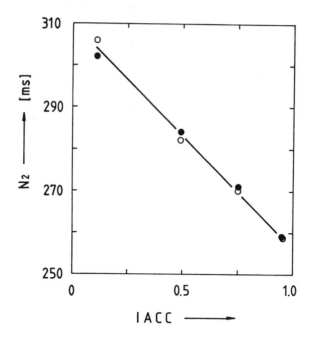

FIGURE 5.21. The linear relationship between the IACC and the N_2-latency, $p < 0.01$ (Ando, Kang, and Nagamatsu, 1987). (●): N_2-latency of SVR over the left hemisphere; (○): N_2-latency of SVR over the right hemisphere; and (————): regression.

5.2.2. Slow Vertex Responses (SVRs)

Previously, four significant and independent physical factors have been discussed consisting of time and space criteria of the sound field in a concert hall. Efforts to describe the important qualities of sound in terms of the processes of the auditory pathways and the brain have been brought to bear on the problem. If enough were known about how the auditory and the central nervous systems modify the nerve impulses from the cochlea, the design of concert halls, for example, could proceed according to guidelines derived from the knowledge of these processes. Attempts to approach this have been made through a study of the auditory evoked potentials over the left and right human cerebral hemispheres.

(A) Recording Slow Vertex Responses as
    Auditory Evoked Potential (AEP)

Prior to the test, each subject was asked to abstain from smoking and from drinking any kind of alcoholic beverage for about 12 hours. In order to compare the results of the SVRs with the subjective preference obtained by paired-comparison tests, a reference stimulus was first presented and then the adjustable test stimuli were presented. Such pairs of stimuli were presented alternately 50 times and the SVRs were recorded. The electrical responses were obtained from the left and right temporal area (T_3 and T_4) according to the International 10–20 System (Jasper, 1958). The reference electrodes were located on the right and left earlobes and were connected together. Figure 5.22 shows examples of the SVR amplitude (as an AEP amplitude), obtained by averaging the 50 responses for a single subject, as a function of the delay time of a single reflection. The amplitude of the reflection was the same as that of the direct sound $A_0 = A_1 = 1$, and the source signal was a fragment of a continuous speech (Japanese) "ZOKI–BAYASHI" (meaning a grove or a copse) of 0.9 s. The reference sound field was only the direct sound, without any time delay, and the total sound pressure levels were kept constant in this experiment. Two loudspeakers producing the direct sound and the single reflection were located together in front of the subject, so that the magnitude of the interaural cross-correlation (IACC) could be kept at a constant value of nearly unity for all sound fields tested here.

From Figure 5.22, we can find the maximum latency at the most preferred delay time of reflection to be 25 ms, indicating a relaxation. This delay time corresponds

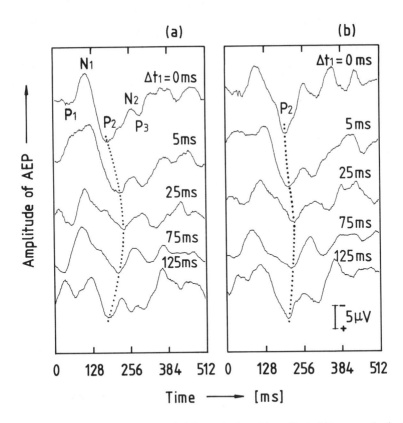

FIGURE 5.22. Averaged SVRs recorded for a single subject. Dotted lines are the loci of P_2 latency for the delay time of the reflection. The upward direction indicates negativity. (a) left hemisphere; and (b) right hemisphere.

to the effective duration of the autocorrelation function (ACF) of the continuous speech signal (Ando, Kang, and Morita, 1987).

(B) Hemispheric Dominance Depending on the Spatial and Temporal Factors

Figure 5.23 shows amplitudes of the early SVR, $A(P_1 - N_1)$, as a function of the delay time of reflection. The values, averaged for eight normal subjects, are plotted in this figure. The source signal was continuous speech. The solid line indicates the amplitude from the left hemisphere and the dashed line is the amplitude from the right hemisphere. Obviously, the amplitude from the left is greater than that from the right ($p < 0.01$). This may indicate the left hemisphere dominance or specialization of the human brain for such a change of the delay time of reflection for speech (see also Table 5.1). When the sensation level was changed, the amplitude of the SVR from the right hemisphere was greater than that from the left hemisphere except for that at 30 dB of the SL, even if the speech signal was used as shown in Figure 5.24 (Nagamatsu, Kasai, and Ando, 1989). When the IACC was changed in the paired stimuli, using 1/3-octave-band noise with the center frequency of 500 Hz, then the amplitude from the right hemisphere was much greater than that of the left as shown in Figure 5.25 (Ando, Kang, and Nagamatsu, 1987). When we put all these data in together, Table 5.1 shows that the hemispheric dominance changes for different sound signals and a change to one of the acoustic factors of the sound fields. It is remarkable that hemispheric dominance appeared

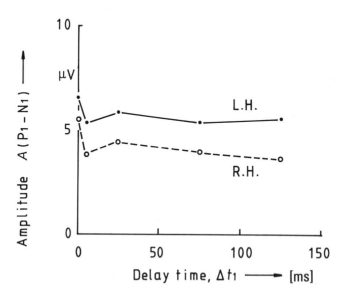

FIGURE 5.23. Averaged amplitudes $A(P_1 - N_1)$ of the test sound field over the left and right hemispheres, as a function of the delay time of the reflection (eight subjects). (————): left hemisphere; and (— — — —): right hemisphere.

TABLE 5.1.   Amplitude differences of the SVR, $A(P_1 - N_1)$ over the left and right cerebral hemispheres. See also Table 5.4.

Source signal	Parameter varied	$A(P_1 - N_1)$	Significance level
Speech (0.9 s)	SL	$R > L$	< 0.01
Speech (0.9 s)	Δt_1	$L > R$	< 0.01
Speech vowel /a/	IACC	$R > L$	< 0.025
1/3 Oct. band noise	IACC	$R > L$	< 0.05

on only the amplitude component of the SVR. It is found here that the right hemisphere was dominant for "the continuous speech" signal under the condition of varying the SL, while the left hemisphere was dominant under the condition of varying the delay time of reflection, which is a temporal criterion of the sound field. If the IACC was changed, then the right hemisphere was highly activated due to the spatial criterion and noise stimulus.

As is well known, the left hemisphere is mainly associated with speech and time-sequential identifications, and the right is concerned with nonverbal and spatial identifications (Kimura, 1973; Sperry, 1974). It is considered here, however, that hemispheric dominance is a relative phenomenon depending on what is changed in

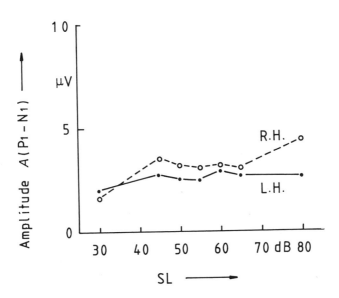

FIGURE 5.24. Averaged amplitudes, $A(P_1 - N_1)$ of the test sound field over the left and right hemispheres, as a function of the sensation level (five subjects). (————): left hemisphere; and (– – – – –): right hemisphere.

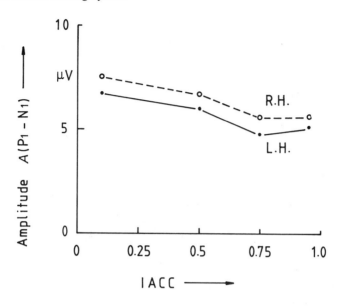

FIGURE 5.25. Averaged amplitudes, $A(P_1 - N_1)$ of the test sound field over the left and right hemispheres, as a function of the IACC (eight subjects). (————): left hemisphere; (------) : right hemisphere.

the comparasion pair, i.e., the temporal criterion or spatial criteria and no absolute behavior could be observed.

(C) Relationship Between the N_2-Latency and Subjective Preference

Figure 5.26 summarizes the relationship between the scale values of subjective preference (subjective diffuseness in the change of the IACC) and the latency components of the SVR, while the behavior of the amplitude component indicated hemispheric dominance as discussed above. Applying the paired method of stimuli, both the SVR and the subjective preference for sound fields were investigated as functions of the sensation level and the time delay of the single reflection. As is mentioned above, the source signal was continuous speech with a 0.9 s duration.

The results of the scale value of subjective preference are indicated in the upper part of Figure 5.26, while the lower part indicates the appearance of latency components. As shown in the left and center columns in this figure, the neural information related to subjective preference appeared typically in an N_2-latency of 250 ms to 300 ms, when the SL and the delay time of the reflection Δt_1 were changed.

Further details of the latencies for both the test sound field and the reference sound field, when Δt_1 was changed, are shown in Figure 5.27. Interestingly, the parallel latencies at P_2, N_2, and P_3, were clearly observed as functions of the delay

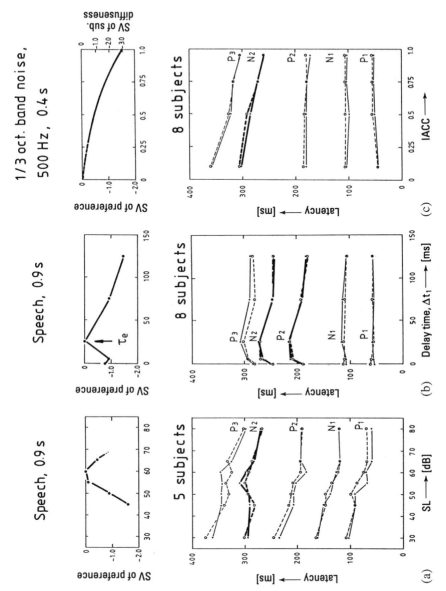

FIGURE 5.26. Relationships between averaged latencies of SVR and subjective preference for three objective parameters. (————): Left hemisphere; (– – – – – –) : Right hemisphere. (a) As a function of the sensation level (SL); (b) as a function of the delay time of reflection, Δt_1; and (c) as a function of the IACC.

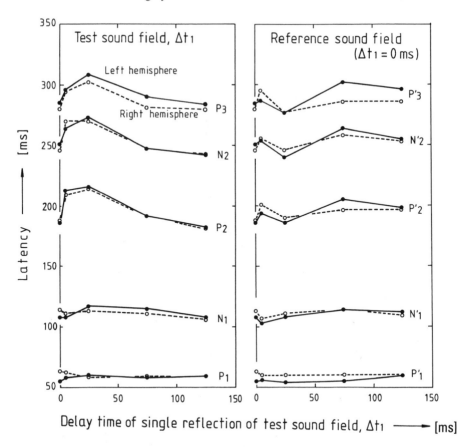

FIGURE 5.27. Averaged latencies for both the test sound field and the reference sound field for paired stimuli, as a function of the delay time of the reflection, Δt_1. (——————): left hemisphere; (– – – – – –) : right hemisphere. Maximum latencies of P_2, N_2, and P_3 are found at $\Delta t_1 = 25$ ms, while relatively short latencies of $P'2$, $N'2$, and $P'3$ are observed.

time Δt_1. However, latencies for the reference sound field ($\Delta t_1 = 0$) in the paired stimuli are found to be relatively shorter, while the latencies for the test sound field ($\Delta t_1 = 25$ ms, the most preferred delay) become longest. This may indicate a kind of contrast process underestimating the reference sound field when it is compared with a preferred sound field.

In general, relatively long-latency responses are observed in the subjectively preferred range of each factor. Thus, the difference of N_2-latencies over both hemispheres in response to a pair of sound fields contains almost the same information obtained from paired-comparison tests for preference, as does primitive subjective response.

The right column of Figure 5.26 shows the effects of varying the IACC using 1/3-octave-band noise (500 Hz). At the upper part, the scale value of the subjective diffuseness is indicated as a function of the IACC. The scale value of the subjective preference also has a similar behavior plotted against the IACC, when speech or music signals are presented as described in the previous section. The information related to subjective diffuseness, therefore, appears in the N_2-latency, ranging from 260 ms to 310 ms, in which a tendency for an increasing latency while decreasing the IACC was observed for eight subjects (except for the left hemisphere of one subject). As already indicated in Figure 5.21, the relationship between the IACC and the N_2-latency was found to be linear and the correlation coefficient between them was $-0.99(p < 0.01)$.

Furthermore, let us look at the behavior of the early latencies of P_1 and N_1. These were almost constant when the delay time and the IACC were changed. However, the information related to the sensation level and loudness may be found typically at the N_1-latency. This tendency agrees well with the result of Botte, Bujas, and Chocholle (1975).

Consequently from 40 ms to 170 ms of the SVR, the hemispheric dominance may be found for the amplitude component, which may be called specialization of the left and right hemispheres. Then the latency differences corresponding to the sensation level may be found in the range of 120 ms to 170 ms. Finally, it is found that the N_2-latency components in the delay range between 200 ms and 310 ms may well correspond to the subjective preference relative to the listening level, the time delay of the reflection, and indirectly the IACC. Since the longest latency was always observed at the most preferred condition, it is concluded that the larger part of the brain may be relaxed at the preferred condition, causing the observed latency behavior to occur.

As discussed in Section 5.2.1, the activity of the ABR in the short delay range (less than 10 ms) after the sound signal has arrived at the eardrums, may indicate a possible mechanism in the auditory pathways for detecting the magnitude of the IACC.

5.2.3. SVR in Change to Both the Initial Time Delay of Reflections and the IACC

The scale values of sound fields estimated in a series of subjective preference judgments, conducted for the purpose of designing concert halls and auditoria, have been well described by four independent factors, including both temporal and spatial ones. The auditory-evoked potentials over the left and right human cerebral hemispheres were analyzed in an effort to model auditory–brain systems for responding to sound fields, and a possible mechanism of the interaural cross-correlation function at the inferior colliculus in the auditory pathways was found by recording the left and right auditory brainstem responses. Analyses of the slow vertex responses (SVR) revealed that the N_2-latency in the delay range between

200 ms and 310 ms corresponds well to the subjective preference with reference to the listening levels, the initial time delay of the reflection, and the IACC. So far, results of the SVR in the change of each single factor of sound have been described.

Now, we have to examine whether or not the initial time delay of reflections as a temporal factor and the IACC as a spatial factor have influence on the N_2-latency independently (Ando and Mizuno, data). The experimental subjects were two right-handed Japanese students, subjects W and U. The SVR was recorded from the left and right temporal areas, respectively, and labeled as T_3 and T_4. A pair of sound fields consisting of the reference stimulus ($\Delta t_1 = 0$ ms and IACC $= 0.99$: the direct sound was reproduced by a loudspeaker in front of the subject) and the test stimulus was presented. The sound-pressure level measured at the ear entrance was held constant at a peak value of 75 dB(A). The sound signal used was a male voice of 0.9 s duration reading the words of the Japanese poem, "ZOUKI–BAYASHI." The silent interval between the sound signals was 1.1 s. The effective duration τ_e of the autocorrelation function (ACF) of the speech signal was 35 ms for the direct sound reproduced by the loudspeaker used here.

To evaluate the independence of the two factors by their influence on the N_2-latency, the initial time-delay gap between the direct sound and the first reflection of the test stimuli was adjusted to be

$$\Delta t_1 = 25, 35, \text{ and } 70 \quad [\text{ms}],$$

and the IACC of the sound field was controlled by interchanging the two directional reflections (Ando and Mizuno, data), so that

$$\text{IACC} = 0.27, 0.49, \text{ and } 0.59.$$

The amplitudes of the two reflections were both -3 dB with respect to that of the direct sound, and the delay time between the two reflections was 5 ms making incoherent reflections. Thus, a total of nine sound fields was used as test stimuli, and the test was repeated five times for each subject.

An example of the averaged N_2-latencies of the SVR, from the left and right hemispheres of subject W is shown in Figure 5.28 as a function of Δt_1, for three IACC values. Similar to the results of the previous sections, significant changes due to the two factors are obtained only in the N_2-latencies. Here, N_2-latencies for subject W depended on both Δt_1 ($p < 0.01$) and IACC ($p < 0.05$) from both hemispheres, and there is no significant interaction between these factors on the N_2-latency. This study confirms the independent influence of both factors, such that the longest N_2-latencies are obtained when $\Delta t_1 = \tau_e$ (35 ms when the total amplitude of reflections $A = 1.0$), which corresponds to the most preferred condition, and the latencies are longer when the IACC are smaller. No significant differences could be achieved between the left and right hemispheres in this experiment, at least due to the changes of temporal and spatial criteria associated with the different hemispheres.

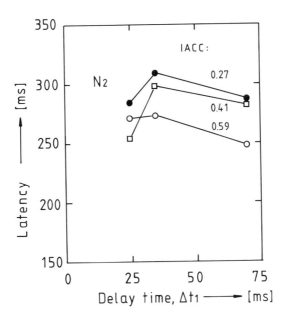

FIGURE 5.28. Latencies of the SVR from the left hemisphere as a function of the initial time delay Δt_1 and as a parameter of the IACC of sound fields for a single subject (Ando and Mizuno, data).

The N_2-latency values for subject U also indicated the maximum at $\Delta t_1 = 35$ ms ($p < 0.01$), but there was no significant differences between IACC $= 0.27$ and 0.59. There are probably two causes for this result:

(1) the small range of the IACC varied, between 0.27 and 0.59; and
(2) the individual differences between subjects, similar to the difference of subjective preference judgments as discussed in Chapter 9.

5.3. Influence of the Continuous Brain Wave (CBW) on Subjective Preference

Thus far, we have discussed results obtained by adding the auditory evoked potentials (SVR) up to 500 ms in the change of the SL, Δt_1, and the IACC, using short signals less than 0.9 s. However, for a wide range of reverberation times (T_{sub}), no useful data could be obtained by the SVR.

The purpose of this section is to find a distinctive feature in the continuous brain wave (CBW) for the T_{sub} with a long signal duration. Before going into detail, a preliminary study was performed in the change of the delay time of a single reflection, to reconfirm the SVR results as discussed in Section 5.2 (Ando and Chen, 1996).

5.3.1. CBW in the Change of the Delay Time of a Single Reflection

In this experiment, Music Motif B (Arnold: Sinfonietta of Opus 48, a 5 s piece of the third movement) was selected as the sound source. The delay time of the single reflection Δt_1 was alternatively adjusted to 35 ms (a preferred condition) and 245 ms (a condition of echo disturbance). The CBW of ten pairs from T_3 and T_4 was recorded for about 140 s in one day, and the experiments were repeated over a total of three days with 11 subjects (male students, 24 ± 2 years of age). In an anechoic chamber, the subject was asked to close his eyes to concentrate on listening to the music during recording of the CBW. The IACC was kept at a constant value of nearly unity. Two loudspeakers in front of the subject were set up. The sound-pressure level was fixed at 70 dBA, in which the amplitude of the single reflection was the same as that of the direct sound, $A_0 = A_1 = 1$. The leading edge of each sound signal was recorded at the same time for analyses of the CBW. The CBW recorded was sampled at least 100 Hz after passing through a filter of $5 - 40$ Hz with a slope of 140 dB/Oct.

In order to analyze a possible brain activity corresponding to the subjective preference, an attempt was made to analyze the effective duration of the ACF, τ_e in the α-wave range (8 Hz to 13 Hz) of the CBW. First of all, considering the fact that the subjective preference judgment needs at least 2 s to develop a "psychological present," the running integration interval $(2T)$ was examined by changing between 1.0 s and 4.0 s. A satisfactory duration $2T$ in the ACF analyses was found only from the left hemisphere for 2–3 s, but not from the right.

Table 5.2 indicates the results of the analysis of variance for values of τ_e in the α-wave obtained at $2T = 2.5$ s. Though the individual difference is great ($p < 0.01$), a significant difference is obtained for Δt_1 (LR: $p < 0.025$); however, a significant difference is also observed for an interference effect between factors Δt_1 and LR ($p < 0.01$). Therefore, in order to analyze the data in more detail for each Δt_1 and LR, we show the averaged value of τ_e in the α-wave with 11 subjects in Figure 5.29. It is clearly shown that the values of τ_e at $\Delta t_1 = 35$ ms are significantly longer than those at $\Delta t_1 = 245$ ms ($p < 0.01$) only on the left hemisphere, but not on the right. Ratios of τ_e values in the α-wave range $\Delta t_1 = 35$ ms to 245 ms for each subject are shown in Figure 5.30. All of the individual data indicate that the ratios in the left hemisphere are much longer than those in the right hemisphere at the preferred condition of 35 ms.

Thus, the results reconfirm that, when Δt_1 is changed, the left hemisphere is greatly activated (Figure 5.23), and the value of τ_e on the α-wave on this hemisphere corresponds well with the subjective preference. The α-wave, which has the longest period in the CBW in the awakening stage and which may indicate a fullness of "pleasantness" and "comfortableness" or a preferred condition, is widely accepted.

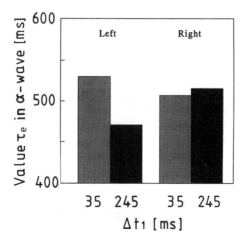

FIGURE 5.29. Averaged value of τ_e in the α-brain-wave range for a change of Δt_1: 35 ms and 245 ms (eleven subjects). Left: left hemisphere; and Right: right hemisphere.

Thus, a large value of τ_e in the α-wave may relate to the large N_2-latency of the SVR at the preferred condition discussed in Section 5.2.

5.3.2. CBW in the Change of the Subsequent Reverberation Time

Now, let us examine values of τ_e in the α-wave for a change of the subsequent reverberation time (T_{sub}) with 10 subjects relative to the scale values of subjective preference (Chen and Ando, 1996). The sound source used was Music Motif B, the same as above. The CBWs from the left and right hemisphere were recorded. Values of τ_e in the α-wave for the duration $2T = 2.5$ s were also analyzed here.

TABLE 5.2.   Results of the analysis of variance for values of τ_e of the ACF in the α-wave, in change of Δt_1.

Factors	F	Significance level
Subject	93.1	< 0.01
Hemisphere, LR	1.0	
Delay time, Δt_1	5.8	< 0.025
Subject and LR	8.9	< 0.01
Subject and Δt_1	0.4	
LR and Δt_1	9.6	< 0.01
Subject, LR and Δt_1	0.4	

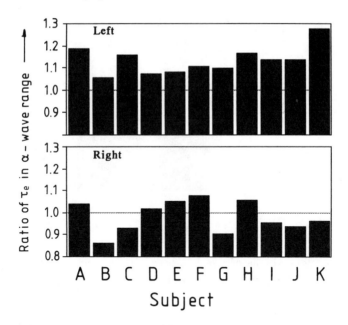

FIGURE 5.30. Ratio of the τ_e values in the α-brain-wave range for a change of Δt_1: [τ_e value at 35 ms]/[τ_e value at 245 ms] for each of eleven subjects, A–K. Left: left hemisphere; Right: right hemisphere.

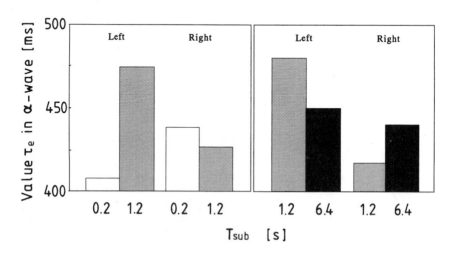

FIGURE 5.31. Averaged value of τ_e in the α-brain-wave range for a change of T_{sub}: 0.2 s and 1.2 s; 1.2 s and 6.4 s (ten subjects). Left: left hemisphere; Right: right hemisphere.

First consider the averaged values of τ_e in the α-wave as shown in Figure 5.31. Clearly, the values of τ_e are much larger at $T_{sub} = 1.2$ s (a preferred condition) than those at $T_{sub} = 0.2$ s in the left hemisphere, while the values of τ_e are larger at $T_{sub} = 1.2$ s (a preferred condition) than those at $T_{sub} = 6.4$ s also in the left hemisphere. However, these facts are not true for the right hemisphere; rather, the contrary is true.

The results of the analysis of variance are indicated in Table 5.3. Although there is a large individual difference, a significant difference is achieved for T_{sub} in the pair 0.2 s and 1.2 s ($p < 0.05$), and interference effects are observed for the factors Subject and LR ($p < 0.01$), and LR and T_{sub} ($p < 0.01$). No such significant differences are achieved for the pair 1.2 s and 6.4 s, but there are interference effects between the Subject and LR, and the Subject and T_{sub}. Thus, in order to discuss the matter in more detail, the ratio of values of τ_e in the α-wave are shown in Figure 5.32 for each subject. All of the individual data indicate that the ratios in the left hemisphere are much larger than those in the right hemisphere at $T_{sub} = 1.2$ s relative to $T_{sub} = 0.2$ s (Figure 5.32(a)). However, this is not the case of $T_{sub} = 1.2$ s relative to $T_{sub} = 6.4$ s, indicating large individual differences (Figure 5.32(b)).

In fact, these individual results correspond well to the scale-values of individual subjective preference. Figure 5.33 shows the scale values of preference as a function of T_{sub} for each subject. The most preferred values of T_{sub}, which were different for each subject, averaged at about 1.2 s. The ratio of the values τ_e in the α-wave at 1.2 s and 6.4 s is well correlated to the difference of the scale-values of the subjective preference of each individual, reflecting a large individual difference, as shown in Figure 5.34 ($r = 0.70$, $p < 0.01$).

The CBW in change of the IACC has been investigated (Nishio and Ando, 1996). Clearly, in change of the IACC using Music Motif B, the right hemisphere dominance is activated by the analyses of the value of τ_e in the α-wave ($p < 0.001$). Then this activity propagates toward the left hemisphere. Such an activity is derived from calculation of the cross-correlation function between the CBWs derived from the different electrodes.

Table 5.4 summarizes the hemisphere dominance obtained by analyses of the values of τ_e in the α-wave on a change of Δt_1, T_{sub}, and the IACC. This conclusion may suggest that the value of τ_e in the α-wave is an objective index for obtaining excellent conditions of human environment so far as the brain is concerned (Section 12.1). Note that no preference information appeared in the amplitude $\Phi(0)$ of the running ACF of α-waves.

5.4. Auditory–Brain System: A Proposed Model

5.4.1. Background

This model is based on the following facts: First of all, we are interested in the fact that the human ear sensitivity to the sound source in front of the listener is

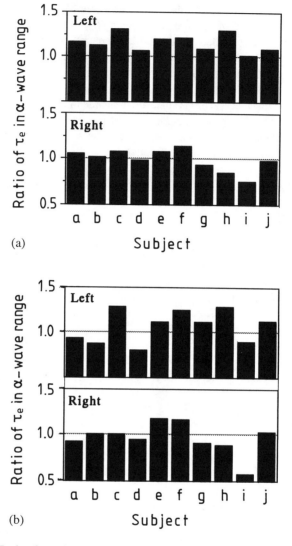

FIGURE 5.32. Ratio of τ_e values in the α-brain-wave range for a change of T_{sub} for each of ten subjects, (a)–(j). (a) [τ_e value at 1.2 s]/[τ_e value at 0.2 s]; and (b) [τ_e value at 1.2 s]/[τ_e value at 6.4 s]. Left: left hemisphere; Right: right hemisphere.

essentially formed by the physical system from the source point to the oval window of the cochlea as discussed in Section 5.1.

By recording the left and right ABRs it has been found that:

(1) Amplitudes of waves I and III correspond roughly to the sound-pressure level as a function of the horizontal angle of incidence to the listener (ξ).

TABLE 5.3. Results of the analysis of variance for values of τ_e of the ACF of the α-wave in change of T_{sub}.

Factor	F	Significance level	F	Significance level
	(Pair of 0.2 s & 1.2 s)		(Pair of 1.2 s & 6.4 s)	
Subject	40.9	< 0.01	40.2	< 0.01
LR	2.1		2.0	
T_{sub}	6.2	< 0.025	0.02	
Subject and LR	2.8	< 0.01	2.0	< 0.05
Subject and T_{sub}	1.2		2.7	< 0.01
LR and T_{sub}	14.0	< 0.01	0.2	
Subject, LR and T_{sub}	1.3		0.7	

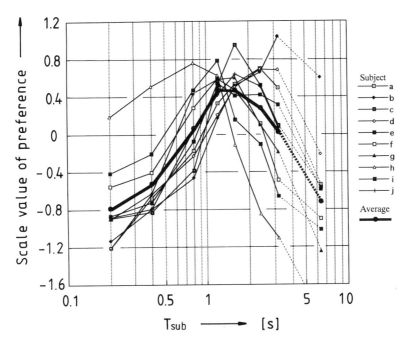

FIGURE 5.33. Scale values of the subjective preference as a function of T_{sub} obtained by the paired-comparison tests for each of ten subjects, (a)–(j). Scale values at 6.4 are extrapolated by the curve of 3/2 power of $\log_{10}(T_{sub})$, according to the results shown in Figure 4.8.

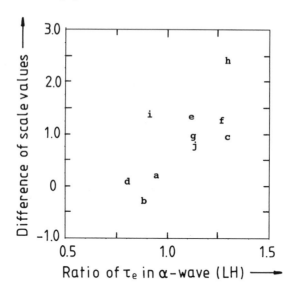

FIGURE 5.34. Relationship between the difference of scale values [SV (1.2 s) to SV (6.4 s)] and the ratio of τ_e values in the α-brain-wave range of the left hemisphere for a change in T_{sub} for each of ten subjects, (a)–(j).

(2) Amplitudes of waves II and IV correspond roughly to the sound-pressure level as a function of the contra horizontal angle $(-\xi)$, implying the interchange of a neural information flow between the left and right hemispheres.
(3) Results of analyses of the ABRs indicate that possible neural activities at the inferior colliculus correspond well to the values of the IACC.

Also, it has been discovered by recording the left and right SVRs that:

(4) The left and right amplitudes of the early SVRs, $A(P_1 - N_1)$, indicate that the left and right hemispheric dominances are due to the temporal factor, Δt_1, and the spatial factors, LL and IACC, respectively, as indicated in Table 5.1.

TABLE 5.4. Hemisphere dominance obtained by analyses of the values of τ_e in the α-wave in change of Δt_1, T_{sub} and the IACC. See also Table 5.1.

Source signal	Parameter varied	Ratio of values of τ_e for α-waves	Significance level
Music Motif B	Δt_1	$L > R$	$p < 0.01$
Music Motif B	T_{sub}	$L > R$	$p < 0.01$
Music Motif B	IACC	$R > L$	$p < 0.01$

(5) Both the left and right latencies of N_2 correspond well to the scale values of subjective preference as a primitive response.

In addition to the above mentioned facts:

(6) Results of the CBW for the cerebral hemispheric specialization of the temporal factors, Δt_1 and T_{sub} indicate the left hemisphere dominance, and the IACC indicates the right hemisphere dominance. Thus, a high degree of independence between the left and right hemispheric factors can be achieved.
(7) Values of τ_e in the α-waves from dominant hemispheres correspond to the scale value of subjective preference.
(8) The results of subjective preference, coloration, and other important subjective attributes are well described in relation to both the autocorrelation function of source signals and the interaural cross-correlation function.

It is worth noticing that the SL or LL is classified as a temporal-monaural factor in the sense of the physical viewpoint. However, the results of the SVR indicate that the SL is the right hemisphere dominance (Section 5.2.2). Thus, hereafter, the SL or LL is classified as a spatial factor, which is also expressed by the geometric average value of sound energies arriving at the two ears, given by Equation (3.24).

Based on these physiological responses, a model of the auditory–brain system may be proposed for the major independent acoustic factors, classified by comprehensive temporal and spatial factors, which are well represented in the model. The model consists of the autocorrelation mechanisms, the interaural cross-correlation mechanism between the two auditory pathways, and the specialization of human cerebral hemispheres for temporal and spatial factors of the sound field.

5.4.2. The Model

According to the relationship of subjective attributes, and the phenomena to the auditory-evoked potentials, including the CBW in the change of acoustic factors, a model can be proposed as shown in Figure 5.35. In this figure, a sound source $p(t)$ is located at r_0 in a three-dimensional space and a listener sitting at r is defined by the location of the center of the head, $h_{l,r}(r|r_0, t)$, being the impulse responses between r_0 and the left and right ear-canal entrances. The impulse responses of the external ear canal and the bone chain are $e_{l,r}(t)$ and $c_{l,r}(t)$, respectively. The velocities of the basilar membrane are expressed by $V_{l,r}(x, \omega)$, x being the position along the membrane.

The action potentials from the hair cells are conducted and transmitted to the cochlear nuclei, the superior olivary complex including the medial superior olive, the lateral superior olive and the trapezoid body, and to the higher level of two cerebral hemispheres as shown in Figure 5.35.

According to the tuning of a single nerve fiber (Katsuki et al., 1958; Kiang, 1965), the input power density spectrum of the cochlea $I(x')$ can roughly be mapped at a certain nerve position x'. This fact may be partially supported by the ABR waves (I–IV) which reflect the sound-pressure levels as a function of the

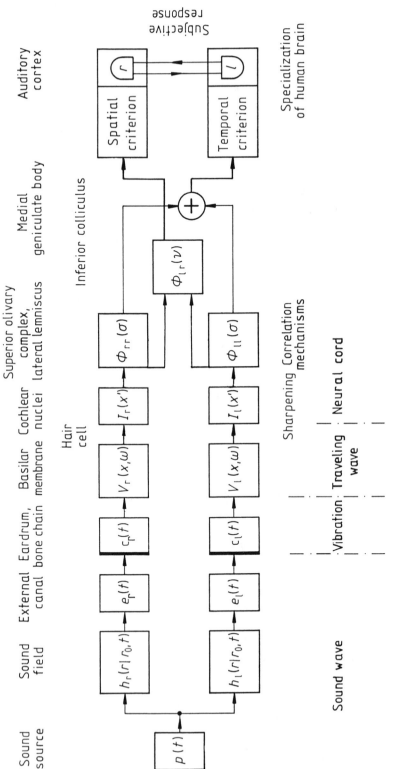

FIGURE 5.35. An auditory–brain model for subjective responses.

horizontal angle of incidence to a listener (Section 5.2.1). Such neural activities, in turn, include sufficient information to attain the ACF at a higher level, probably near the lateral lemniscus, as indicated by $\Phi_{ll}(\sigma)$ and $\Phi_{rr}(\sigma)$. The interchange of neural signals discussed in Section 5.2.1 is not included here for convenience. As is also discussed in Section 5.2.1, the neural activity (wave V) may correspond to the IACC. Thus, the interaural cross-correlation mechanism may exist at the inferior colliculus. It is concluded that the output signal of the interaural cross-correlation mechanism including the IACC and the loci of maxima may be dominantly con-nected to the right hemisphere. Also, the sound-pressure level, which corresponds to the denominator of Equation (3.23) with the ACFs for the two ears at the origin of time ($\sigma = 0$) which, in fact appears in the latency at the inferior colliculus, as shown in Figure 5.20, may be processed in the right hemisphere. Effects of the initial time-delay gap between the direct sound and the single reflection Δt_1 included in the autocorrelation function may activate the left hemisphere. The spe-cialization of the human cerebral hemisphere may relate to the highly independent contribution between the spatial and temporal criteria to subjective attributes. It is remarkable that, for example, "cocktail party effects" might well be explained by such specialization of the human brain, because speech is processed in the left hemisphere, and spatial information is mainly processed in the right hemisphere.

5.4.3. Subjective Responses from the Model

So-called "over-all responses," such as subjective preference, must be associated with both hemispheres and with all four acoustic factors of the temporal and spatial variables. As discussed in Section 6.4, speech intelligibility and clarity may be influenced by all four factors including the IACC (Nakajima and Ando, 1991; Nakajima, 1992) associated with both human cerebral hemispheres. In general, any subjective attributes is described by all the acoustic factors associated with both cerebral hemispheres.

For example, the apparent source width (ASW) is considered to depend on the IACC and the listening level (LL), as previously reported by Keet (1968) (see also Section 6.1). However, the subjective diffuseness may to some extent be related to Δt_1 and T_{sub} under the fixed conditions of the IACC and LL. Also, echo disturbance, the threshold of perception of a reflection and coloration (temporal information) are considered to be dominantly processed in the left hemisphere related to the envelope function of the ACF of the source signals and Δt_1 (Ando, 1986), but they may depend more or less on the IACC as well. Furthermore, the IACC as a subjective diffuseness factor is dependent on Δt_1 for a very short delay range $0 < \Delta t_1 < 0.1\tau_e$, increasing the value of the IACC (Ando, 1977).

Any other subjective responses may well be described in such a manner by at least four orthogonal factors, together with the ACF of the source signal and the total amplitude of reflections, similar to the analysis of subjective preference, as is discussed in the following chapter.

6

Important Subjective Attributes for the Sound Field, Based on the Model

Based on the model of the previous chapter, we can describe qualities of sound fields in terms of processes of the auditory pathways and the brain. The power density spectra in the neural activities in the left and right auditory pathways have a sharpening effect (Katsuki et al., 1958; Kiang, 1965). This information is enough to attain approximately the autocorrelation functions $\Phi_{ll}(\sigma)$ and $\Phi_{rr}(\sigma)$, respectively, where σ corresponds to the neural activities. Together with the mechanism of the interaural cross-correlation function found in Section 5.2, fundamental subjective attributes may then be well described.

6.1. Subjective Diffuseness and ASW in Relation to the IACC and/or the W_{IACC}

The interaural cross-correlation function is a significant factor in determining the perceived horizontal direction of a sound and the degree of subjective diffuseness of a sound field (Damaske and Ando, 1972). A well-defined direction is perceived when the normalized interaural cross-correlation function has one sharp maximum (a small value of W_{IACC} defined in Figure 3.7). On the other hand, subjective diffuseness or no spatial impression corresponds to a low value of the IACC ($<$ 0.15). The subjective diffuseness or spatial impression of the sound field in a room is one of the fundamental attributes in describing good acoustics. If the sound arriving at the two ears are dissimilar (IACC $= 0$), then the different signals (but containing the same information) are conveyed through two channels of the auditory system to the brain. This condition, in turn, improves speech clarity as is discussed in Section 6.4.

In order to obtain the scale value of subjective diffuseness, paired-comparison tests with bandpass Gaussian noise, varying the horizontal angle of two symmetric reflections, have been conducted (Ando and Kurihara, 1986; Singh, Kurihara, and Ando, 1994). Listeners judged which of two sound fields were perceived as more diffuse. A remarkable finding is that the scale values of subjective diffuseness are inversely proportional to the IACC, and may be formulated in terms of the 3/2 power of the IACC in a manner similar to the subjective preference values (see Section 4.3), i.e.,

$$S \approx -\alpha(IACC)^\beta, \tag{6.1}$$

where $\alpha = 2.9$, $\beta = 3/2$.

The results of scale values by the paired-comparison test, and the calculated values by Equation (6.1) as a function of the IACC, are shown in Figure 6.1. There is a great variation of data in the range of IACC $<$ 0.5; however, no essential difference may be found in the results with frequencies between 250 Hz and 4 kHz. The scale values of subjective diffuseness, which depend on the horizontal angle, are shown in Figure 6.2, for 1/3 octave-bandpass noise with the center frequencies of 250 Hz, 500 Hz, 1 kHz, 2 kHz, and 4 kHz. Obviously, the most effective horizontal angles of reflections are different depending on the frequency range, and are inversely related to the behavior of the IACC values. These are about $\pm 90°$ for the 500 Hz range and the frequency range below 500 Hz, around $\pm 55°$ for the 1 kHz range, and smaller than for the 2 kHz and 4 kHz ranges (Figure 6.3). The control of directional reflections, for each frequency range, by means of wall-surface structures, is described in Chapter 8.

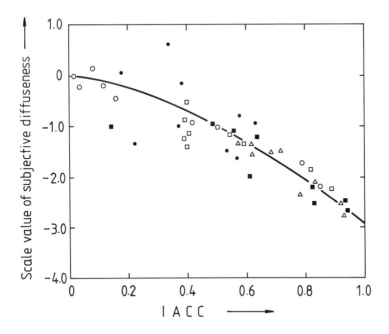

FIGURE 6.1. Scale values of subjective diffuseness as a function of the IACC calculated. Different symbols indicate different frequencies of the 1/3 octave bandpass noise: ($\triangle$): 250 Hz, ($\circ$): 500 Hz, ($\square$): 1 kHz, ($\bullet$): 2 kHz, ($\blacksquare$); and 4 kHz. (————): Regression line by Equation (6.1).

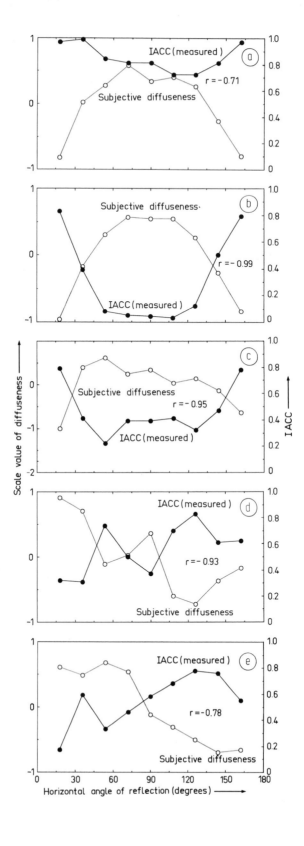

Horizontal angle of reflection (degrees) ⟶

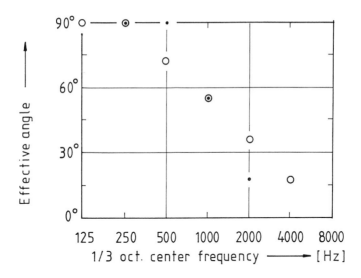

FIGURE 6.3. Effective horizontal angles to a listener to bring about a maximum decrease in the IACC and an increase in subjective diffuseness for the frequency band. (○): Calculated (IACC); and (●): observed (IACC and subjective diffuseness, Figure 6.2).

For a sound field with a predominantly low-frequency range (below 250 Hz), the interaural cross-correlation function has no sharp peaks for the delay range of $|\tau| \leq 1$ ms, and W_{IACC} becomes wider, as is shown in Figure 6.4(a, b). The values shown in Figure 6.4(b) are calculated theoretically by Equation (3.29) with, for simplicity, $\delta = 0.1$ and IACC $= 1$. It is worth noticing that the value of δ may be expressed as a function of the IACC.

Of particular interest is that a wider ASW may be perceived within the low-frequency bands and by decreasing the IACC, as indicated by the equal-ASW curves (Hidaka, Beranek, and Okano, 1995). More clearly, the ASW may well be described by both factors, W_{IACC} and IACC (Sato and Ando, 1996, in which the W_{IACC} defined by the time width of the interaural cross-correlation function crossing zero, for practical convenience; see also Sato and Ando, 1997). Such a perception becomes much more significant, if listeners close their eyes or are listening to loudspeaker reproduction. However, the ASW becomes a minor effect in a real concert hall due to visual perception of the location of the sound sources.

FIGURE 6.2. Scale-values of subjective diffuseness and the IACC as a function of the horizontal angle of incidence to a listener, with 1/3 octave band noise of center frequencies. (a) 250 Hz; (b) 500 Hz; (c) 1 kHz; (d) 2 kHz; and (e) 4 kHz.

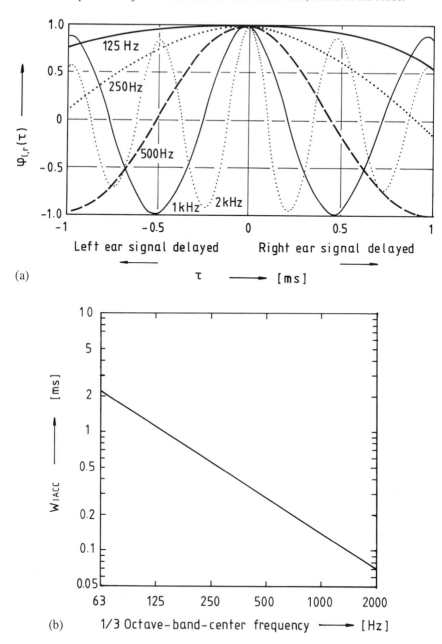

FIGURE 6.4. (a) Measured interaural cross-correlation functions for 1/3 octave-bandpass noise with center frequencies of 125 Hz to 2 kHz; and (b) W_{IACC} as a function of the center frequency.

6.2. Subjective Attributes of the Sound Fields with a Single Reflection in Relation to the ACF of the Source Signals

6.2.1. Preferred Delay of a Single Reflection for Listeners

Results of the subjective preference tests, as an overall psychological response for sound fields with a single reflection, indicated that the most preferred delay of the reflection may be found by the envelope curve of the ACF, defined by the delay τ_p as Equations (4.3) and (4.4),

$$[\Delta t_1]_p = \tau_p,$$

such that

$$|\phi(\tau)|_{\text{envelope}} = kA_1^c, \qquad \text{at} \quad \tau = \tau_p, \tag{6.2}$$

where A_1 is the pressure amplitude of the single reflection, $k = 0.1$ and $c = 1$. If the envelope of the ACF is exponential, then the above equation is simply expressed by Equation 4.1.

Such a relationship also holds for other important subjective responses in relation to the temporal factors discussed below. The constants k and c used in calculating important subjective responses to the sound field, based on the ACF of the source signals, and the amplitude ranges in the experiments, are listed in Table 6.1.

6.2.2. Preferred Frequency Characteristics of Reverberation Time

This section discusses the preferred condition for the reverberation time beyond the treatment of Section 4.3.3. First, in order to obtain a flat frequency response below 3 kHz of the source signals, the right channel of the original recorded signal (Burd, 1969) was mixed with the left channel signal as indicated by the symbol of $(L + R)$ for the Music Motifs B and C (Table 3.1). In controlling the frequency characteristics and eliminating other factors' effects on preference judgments, the following five conditions were imposed (Ando, Okano, and Takezoe, 1989):

(1) the total amplitude of reflection A was adjusted to a constant 4.0;
(2) the listening level was adjusted to 72 dBA or 74 dBA, according to the Music Motif;
(3) the IACC was fixed at nearly 1.0 by the location of the loudspeakers on the median plane;
(4) the delay time of early reflections was fixed at $\Delta t_n = 18.0, 31.1$, and 38.4 [ms], $n = 1, 2, 3$, respectively; and
(5) in order to keep independent reverberations for higher and lower frequency ranges separated by 500 Hz, the time difference between these two signals, produced by the Schroeder Reverberator (modified), was set at 5 ms.

TABLE 6.1. Constants related to the ACF-envelope of source signals for calculating various subjective responses to sound fields with a single reflection, in relation to the ACF-envelope.

Subjective attributes	In Equation (6.1)		Delay time to be obtained	Range of amplitude examined [dB]	Source signals	Authors investigated
	k	c				
Preference of listeners	0.1	1	Preferred delay time	$7.5 \geq A_1 \geq -7.5$	Speech and music	Ando (1977)
Threshold of perception of reflection	2	1	Critical delay time	$-10 \geq A_1 \geq -50.0$	Speech	Seraphim (1961)
50% echo disturbance	0.01	4	Disturbed delay time	$0 \geq A_1 \geq -6.0$	Speech	Haas (1951); Ando et al. (1974)
Coloration	$10^{-2.5}$	-2	Critical delay time	$-7.0 \geq A_1 \geq -27.0$	Gaussian noise	Ando and Alrutz (1982)
Preference of musicians (alto-recorder)	2/3	1/4	Preferred delay time	$-34.0 \geq A_1 \geq -10.0$	Music	Nakayama (1984)

In addition, in order to obtain natural and colorless reverberation, signal-processing filters were set with the following values:

(a) delay times of six comb filters used in the digital reverberator were fixed at the appropriate values for the music motif, i.e., $\tau_i = 28.6, 32.4, 35.3, 37.9, 40.5, 45.8$ [ms] $i = 1, 2, \ldots, 6$, for Music Motif $B(L + R)$, and $\tau_i = 38.1, 43.1, 47.1, 50.5, 53.9, 61.0$ [ms] for Music Motif $C(L + R)$; and
(b) to eliminate the effects of coloration, the delay times of two of the all-pass filters were adjusted within $0.1\tau_e$.

The results of the scale value of preference are shown in Figure 6.5. In this figure, closed circles indicate values of the preferred reverberation time calculated by Equation (4.7) in Section 4.3, with the values $(\tau_e)_{min}$ obtained by the running ACF, $2T = 2$ s, and the running interval of 100 ms through the A-weighting network. The closed squares indicate values calculated by the same equation, with values of $(\tau_e)_{min}$ obtained for two frequency ranges above and below 500 Hz, respectively, after passing through the A-weighting network.

The closed circles are closer to the preferred conditions for both types of music than the closed squares. The reverberation time for the high-frequency range is much more critical than that of the low-frequency range. The range of acceptable values for reverberation time in the low-frequency range is quite wide, from 0.5 to 2.0 times that of the calculated preferred reverberation time (closed circles). Thus, flat frequency responses are in the range of the preferred condition.

6.2.3. Coloration of a Single Reflection

When we listen to sound very near a boundary wall in a room, coloration is clearly perceived. Here, such coloration is discussed in relation to the ACF-envelope (Ando and Alrutz, 1982).

As a source signal, continuous bandpass noise was used, because the ACF of the signal is independent of the time interval extracted for subjective judgments and is theoretically calculable. The normalized ACF of the Gaussian noise after passing through an ideal bandpass filter with a flat response between frequencies f_1 and f_2, $f_2 > f_1$, is given by

$$\phi(\tau) = \frac{2}{\Delta\omega\tau} \sin\left(\frac{\Delta\omega\tau}{2}\right) \cos\left(\frac{\Delta\omega_c\tau}{2}\right), \tag{6.3}$$

where $\Delta\omega = 2\pi(f_2 - f_1)$ and $\Delta\omega_c = 2\pi(f_2 + f_1)$.
The envelope of the ACF is expressed by

$$\frac{2}{\Delta\omega\tau} \sin\left(\frac{\Delta\omega\tau}{2}\right) \qquad \text{for} \quad 0 \le \Delta\omega\tau \le \pi$$

and

$$\frac{2}{\Delta\omega\tau} \qquad \text{for} \quad \Delta\omega\tau > \pi. \tag{6.4}$$

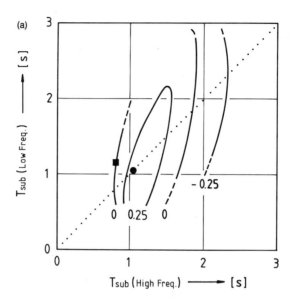

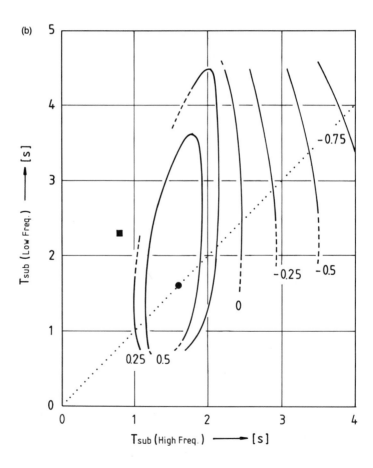

Figure 6.6 shows an example of the measured ACF of the bandpass noise re-produced by a loudspeaker in an anechoic chamber and its envelope curve, which is calculated by putting the equivalent bandwidth for $\Delta\omega$ into Equation (6.4). Since nonideal characteristics of the filter and the loudspeaker were used, the equivalent bandwidth had to be chosen to be greater than the difference of the cut-off frequencies in Equation (6.4). A loudspeaker arrangement that presents the primary sound and the delayed sound is shown on the right side of Figure 6.7. The total sound

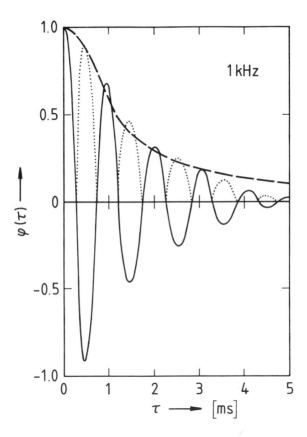

FIGURE 6.6. Measured and calculated ACF of the bandpass noise with a center frequency of 1 kHz. (————): Measured ACF; (· · · · ·): absolute values of the measured ACF; and (— — — —): envelope curve by Equation (6.4).

FIGURE 6.5. Contour lines of equal scale value of the preference for sound fields with two early reflections (fixed) and the subsequent reverberation times, versus T_{sub} (< 500 Hz) and T_{sub} (> 500 Hz). (●): Calculated preferred values at $(\tau_e)_{min}$ for overall frequency range as indicated in Table 3.1; and (■): calculated preferred values at $(\tau_e)_{min}$ for the two frequency ranges. (a) Music Motif $B(L + R)$, SPL = 72 dBA (12 subjects); and (b) Music Motif $C(L + R)$, SPL = 74 dBA (6 subjects).

pressure of the two sounds was automatically kept constant at the listener's position in the anechoic chamber (60 dBA). The subject adjusted the sound-pressure level of the weak delayed sound and judged the threshold of the just noticeable difference which appeared as a coloration, in comparison to the situation in which only the primary sound was presented.

The threshold levels of the weak sound relative to the primary sound are shown in Figure 6.7 as a function of the delay time Δt_1, for a noise source with a center frequency of 1 kHz. The dashed curve represents the calculated values obtained by Equation (6.2) with the envelope of the ACF (Figure 6.6) and using the derived constants, i.e., $k = 10^{-2.5}$ and $c = -2.0$. Similar results were obtained with 250 Hz and 4 kHz, even if the direction of weak sound was changed to $\xi = 36°$ or 90° (Ando and Alrutz, 1982).

6.2.4. Threshold of Perception of a Single Reflection

Seraphim (1961) investigated the perceptibility (aWs) of a single reflection with speech sound as shown in Figure 6.8. The aWs data were obtained under the condition of a single reflection with a horizontal angle to the listeners of $\xi = 30°$. Unfortunately, the ACF of the speech signal used at that time could not be directly related to the behavior of the aWs. However, it was assumed that the ACF of any continuous speech signals with normal speed of speech does not differ much. Let us apply a typical ACF-envelope function of a speech signal as shown in Figure 6.9. Then, the aWs may be described approximately with the ACF-envelope as indicated in the lower part of Figure 6.10.

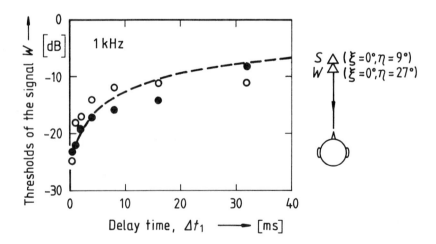

FIGURE 6.7. Threshold level of a weak sound W as a function of the delay time Δt_1 for a bandpass noise with a center frequency of 1 kHz. Different symbols indicate responses with two subjects. (– – – –): calculated values by Equation (6.2) with $k = 10^{-2.5}$, $c = -2$.

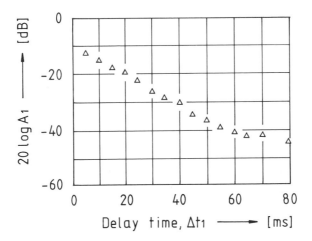

FIGURE 6.8. Threshold level of a single reflection as a function of the delay time Δt_1 for continuous speech signal (Seraphim, 1961).

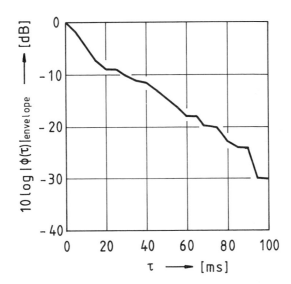

FIGURE 6.9. ACF-envelope function of a typical continuous speech signal.

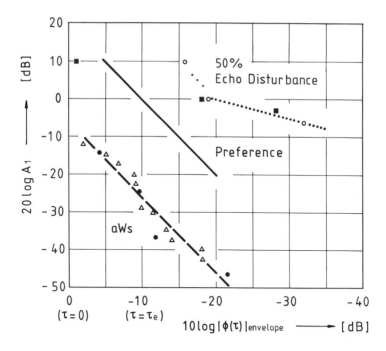

FIGURE 6.10. Amplitude of a single reflection obtained at several subjective responses in relation to the ACF-envelope of music and speech signals. (————): Most preferred condition for the listener calculated by Equation (6.2) with $k = 0.1$, $c = 1$; (————): threshold of perception of the reflection calculated by Equation (6.2) with $k = 2$, $c = 1$; ($\cdots\cdots$): 50% echo disturbance calculated by Equation (6.2) with $k = 0.01$, $c = 4$ for $0 \geq A_1 \geq -6.0$ [dB]; ($\triangle$): aWs by the Beurtleilungsverfahre (Seraphim, 1961) with the ACF-envelope as shown in Figure 6.9; ($\bullet$): Threshold of perception by the limit method (Morimoto et al., 1982) with the ACF-envelope in Figure 6.9.; and ($\circ$,$\blacksquare$): 50% echo disturbances, after Haas (1951) and Ando et al. (1974), respectively, with the ACF-envelope in Figure 6.9.

In order to confirm this result, the values of the perception threshold obtained by the limit method (Morimoto et al., 1982) are plotted in Figure 6.10, where the continuous speech signal with the ACF-envelope shown in Figure 6.9 was also used. Similar results were obtained in spite of the different speech signals used. Such a threshold of perception may be described by Equation (6.2) with $k = 2$ and $c = 1$.

6.2.5. Echo Disturbance of a Single Reflection

In a manner similar to that mentioned above, the echo disturbance data by Haas (1951) and Ando, Shidara, and Maekawa (1974, Stereo-System in Fig. 2), may be rearranged with the ACF-envelope. Results of the 50% echo disturbance are shown in the upper part of Figure 6.10.

Since the echo disturbance effects in the short-delay range of the single reflection within 50 ms are unclear, only the constants in Equation (6.2), for the delay range longer than 50 ms ($10 \log |\phi(\tau)|_{\text{envelope}} < -20$ dB), are meaningful in obtaining $k = 0.01$ and $c = 4.0$. The subjective preferences for listeners are located between the curves of the threshold (aWs) and of the echo disturbance.

6.2.6. Preferred Delay of a Single Reflection for a Performer

From preference judgments with respect to the ease of music performance by alto-recorder soloists (Nakayama, 1984), the most preferred delay time of the single reflection may also be described by Equation (6.2). In this case, the coefficients are $k = 2/3$ and $c = 1/4$.

The coefficient k differs by a factor of about seven from that of the listeners. This indicates that the amplitude of the reflection is evaluated about seven times greater than in the case of listeners, namely, the "missing reflection for performers." Weaker amplitudes are effective and preferred for musicians rather than those preferred by listeners.

Some fundamental subjective attributes have been discussed in relation to the ACF. When the ACF-envelope is expressed approximately by an exponential, then [as is expressed by Equation (4.5)], the corresponding amplitudes of reflection may be described by the normalized delay time of reflection Δt_1 and by the effective duration of the ACF τ_e, as shown in Figure 6.11. For listeners, if $\Delta t_1/\tau_e = 1$, then $20 \log A_1 = 0$ dB. In this figure, the musician's preference of the reflection as ease of performance for playing an alto-recorder is also plotted.

6.3. Loudness in Relation to the Effective Duration of the ACF

Under the fixed conditions of the sound-pressure level (74 dB) and of the other temporal and spatial factors, loudness judgments were performed by changing the ACF of the bandpass noise within the critical band (Merthayasa and Ando, 1996). The effective duration of the ACF, τ_e, or the repetitive feature, of the bandpass noise of 1 kHz center frequency is controlled by the bandpass filter slope used (48, 140 and these in combination by use of two digital recorders, obtaining 1080 dB/octave). The duration of stimuli was 3 s with a rise and fall time of 250 ms, and the interval between stimuli was 1 s. In this test, the bandwidth ΔF of stimuli defined by the -3 dB attenuation of the low and high cut-off frequency was kept constant. In fact, ΔF was set at "0 Hz" with only the slope components controlling the wide range of $\tau_e (= 3.5$–52.6 ms).

The paired-comparison method for judgments was applied to more than six students of normal hearing ability. Since the subjects sat in an anechoic chamber

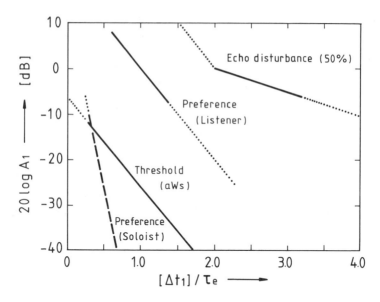

FIGURE 6.11. Amplitude of a single reflection obtained at several subjective responses including the preference of alto-recorder performers as a function of the normalized delay time Δt_1 (see Figure 7.3) by the value of τ_e of music and speech. When the ACF-envelope is exponential, these values may be calculated approximately by Equation (4.5) with constants, k and c (Table 6.1).

facing the loudspeaker located in front at 90 ± 1 cm, the IACC was kept constant at nearly unity.

The scale values of loudness as a function of $\log \tau_e$ are shown in Figure 6.12 (Merthayasa, Hemmi, and Ando, 1994). Obviously, the loudness is influenced by the increasing value of τ_e. Statistical analysis with different values of τ_e indicates a significant difference in loudness ($p < 0.01$). It can be demonstrated that the degree of the repetitive feature of stimuli contributes to the loudness. Since there is no significant difference in loudness between the pure tone and the "0 Hz" bandwidth signal, produced by use of a filter of 1080 dB/octave slopes, use is recommended of a sharp slope-filter in the hearing experiment, which corresponds to the sharpness of the filter in the auditory system (Section 5.1.4).

Furthermore, as shown in Figure 6.13, it is found that the loudness of the band-pass noise within the critical band is not constant. Rather a minimum is indicated at a certain bandwidth, when the filter slope of 1080 dB/octave is used. This evidence differs from the results of Zwicker et al. (1957).

A similar tendency is observed in that, as the reverberation increases, the value of τ_e also increases, as is shown in Figure 6.14(a). Accordingly, loudness increases as the reverberation time increases as shown in Figure 6.14(b) (Ono and Ando, 1996).

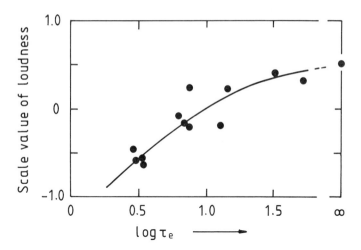

FIGURE 6.12. Scale-value of loudness as a function of log τ_e, in which values of τ_e are measured in ms.

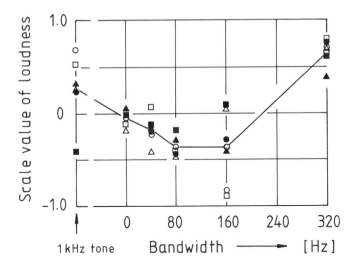

FIGURE 6.13. Scale-values of loudness of bandpass-noise as a function of its bandwidth centered on 1 kHz. The cutoff slope of the filter used was 1080 dB/octave. Different symbols indicate the scale-values obtained with different subjects (six subjects).

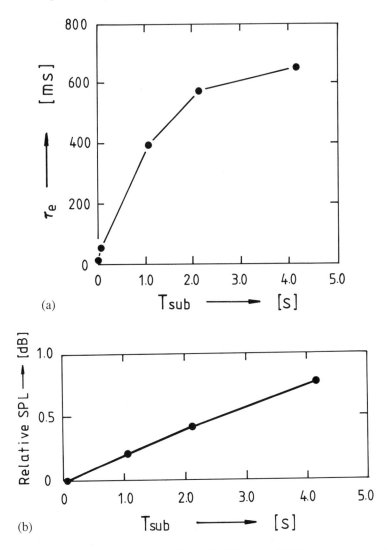

FIGURE 6.14. (a) Effective duration of the ACF of the sound field as a function of the subsequent reverberation time; and (b) the average value of loudness obtained by the constant method as a function of the reverberation time of the sound field.

It is worth noticing that the loudness does not depend on the IACC under the condition of a fixed sound-pressure level at both ear entrances (Merthayasa, Ando, and Nagatani, 1995). This confirms results using headphone reproduction (Chernyak and Dubrovsky, 1968; Dubrovskii and Chernyak, 1969).

6.4. Speech Intelligibility and Clarity in Relation to the Temporal Factor (TF) and the IACC

As far as speech intelligibility is concerned, a speech transmission index (STI) has been proposed by Houtgast, Steeneken, and Plomp (1980) in which the temporal and monaural factors have been taken into consideration. In addition to such monaural effects, binaural effects may contribute to speech intelligibility as well as to speech clarity. For the purpose of examining the effects of the spatial factor on speech identification, speech intelligibility tests were conducted for a synthesized sound field changing the delay time of the single reflection under a constant STI condition (Nakajima and Ando, 1991). In order to obtain the suitable dynamic range of speech intelligibility and to examine effects of both temporal and spatial factors, each monosyllable was joined by both meaningless forward and backward noise maskers instead of continuous speech.

Results are shown in Figure 6.15 which indicates that the speech intelligibility is increased with an increase in the horizontal angle of the single reflection to the listener and with a decrease in the delay time. In order to explain the results, including the spatial effect, there are two simple and comprehensible models to be examined.

(1) The first is the IACC model as described by

$$SI = f(t) + f(s), \tag{6.5}$$

where $t = $ STI and $s = $ IACC. From experimental results, we obtained the relations

$$f(t) \approx 111t + 15t^2 + 50t^3 + 18$$

and

$$f(s) \approx 10.5\,s - 2.5.$$

As shown in Figure 6.16, SI scores calculated by Equation (6.5) agree well with measured ones.

(2) Next, we can introduce BESTI, which is defined by

$$BESTI = \{STI(L), STI(R)\}_{max}, \tag{6.6}$$

where STI(L) or STI(R) indicate the STI at the two ears. As shown in Figure 6.17, the measured SI scores may also be expressed by means of BESTI.

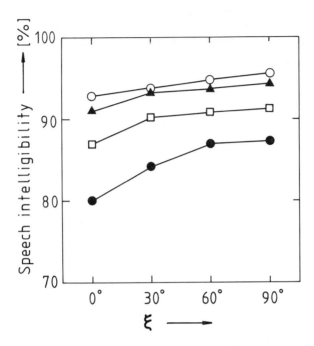

FIGURE 6.15. Speech intelligibility for sound fields with single reflection as a function of the horizontal angle to a listener and as a parameter of the delay time (six subjects). Different symbols indicate different delay times. (○): 15 ms; (△): 30 ms; (□): 50 ms; and (●): 100 ms.

Now the question arises as to which of the two models is more closely related to speech intelligibility for the sound field. A further experiment has been carried out for clarity judgments which are associated with speech intelligibility under the available conditions of varied IACC and fixed BESTI (Nakajima, 1992). On the other hand, conditions of fixed IACC and varied BESTI are physically unreal for the usual sound field in a room, so that an experiment may not be performed under these conditions.

As shown in Figure 6.18, the scale values of clarity, which were obtained by the paired-comparison tests (eight subjects), increases with decreasing IACC for fixed BESTI or STI. The change of the IACC is significant ($p < 0.025$). Therefore, the scale value of speech clarity in the present experimental conditions is expressed by Equation (6.5) with two independent factors, i.e., the IACC as a spatial factor and the STI as a temporal factor, because no significant interference effects between the two factors are observed. A promising method of calculating the speech intelligibility in relation to the delay time of single reflection has been discussed based on the ACF mechanism in the model of the auditory–brain system, as described in Section 5.4. One remarkable finding is that the four factors extracted from the running ACF of source signals and sound fields, i.e., (1) the value of τ_e, (2) the

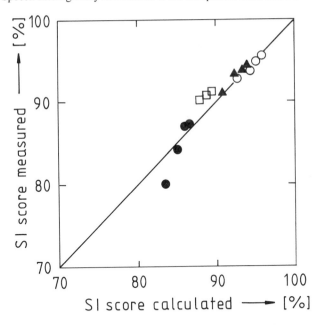

FIGURE 6.16. Relationship between the calculated SI scores by Equation (6.5) and measured scores.

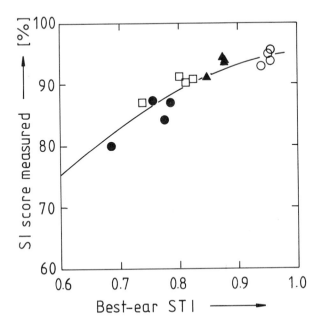

FIGURE 6.17. The SI scores measured in relation to the BESTI, defined by Equation (6.6).

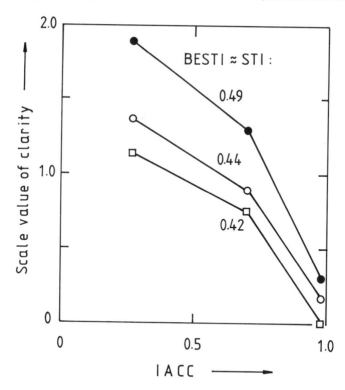

FIGURE 6.18. Scale values of clarity for continuous speech as a function of the IACC and as a parameter of STI. The STI is almost the same as the BESTI in this experimental condition with symmetrical loudspeaker arrangements for the reflections, but the speech clarity differed greatly due to the effect of the IACC (Nakajima, 1992).

delay time of the first peak τ_1, (3) the amplitude of the first peak ϕ_1, and the power of the signal frame $\Phi(0)$, play an important role in recognition of speech (Shoda and Ando, 1996; 1998).

7

Subjective Effects of Sound Field on Performers

The theory of subjective preference allows a conductor or music director to chose a program of music that will be the best sound in a given concert hall. Previously, we have described how music of rapid movement with a short τ_e fits a concert hall with a short initial time-delay gap and a short subsequent reverberation time (Figure 7.7). Music of slow tempo with a long τ_e fuses in a hall with relatively long values for these two temporal factors. It is strongly recommended, therefore, that we choose music motifs to be performed in a given concert hall to be blended with the music program and with the sound field.

7.1. Subjective Preference of Performer for Sound Field on the Stage

As is discussed in Chapter 4, the preferred conditions for listeners are strongly dependent on the effective duration of the autocorrelation function of source signals. Thus, it is assumed that the preferred delay time of reflection for performers on the stage depends on the different types of music sources. In order to support musicians by the stage reflections, Nakayama et al. (1984, 1986, 1988, 1988) examined the preferred conditions of the sound field on the stage by use of seven alto-recorder soloists.

The experiments were conducted by use of a system with delay lines and attenuators as shown in Figure 7.1. The single or two early reflections were simulated in an anechoic chamber by loudspeakers located at a distance of 1.7 m from the center of the head of subject (1.2 m from the floor). The adjustable factors were:

(1) the autocorrelation function with two different music programs as shown in Figure 7.2. These two music selections, with tempos ($\downarrow = 90$ and $\downarrow = 60$) and with different values of τ_e, were composed by Tsuneko Okamoto for this investigation (Table 3.1);

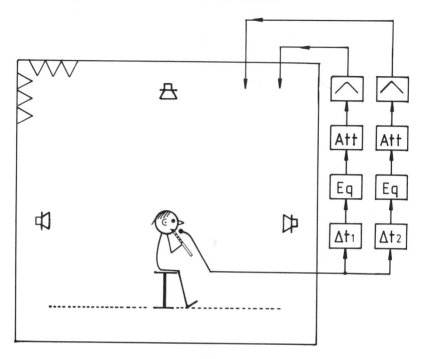

FIGURE 7.1. A diagram of the simulation system for a performer.

(2) the amplitude of reflections relative to that of the direct sound, which is measured at the ear-canal entrances, when the reflection is arriving from the frontal
direction ($\xi = 0°$); and
(3) the angle of incidence of reflections, ξ.

In the first study, adjusting the delay time of a single reflection ($\xi = 0°$) with
fixed amplitudes, eight subjects were asked to respond to the most preferred delay
time $[\Delta t'_1]_p$. The dashed symbol signifies the physical factors for the condition of
the performers due to the different definition of the amplitude of reflection. Results
of the most preferred delay time for a single reflection are shown in Figure
7.3. Clearly, the most preferred delay time for the single reflection differs significantly for the two music motifs played in accordance with the ACF τ_e, and
increases with decreasing amplitude of reflection. For other angles of ξ, if the

FIGURE 7.2. Music scores composed by Tsuneko Okamoto for this experiment and the
measured normalized ACF of the two music motifs, $2T = 32$ s (Nakayama, 1984). (a) Music
scores of Motifs F and G; and (b) normalized ACFs of Music Motif F, $\tau_e = 105$ ms and
Motif G, $\tau_e = 145$ ms. The values of τ_e are obtained by the extrapolation of the first
important decay rate for 5 dB.

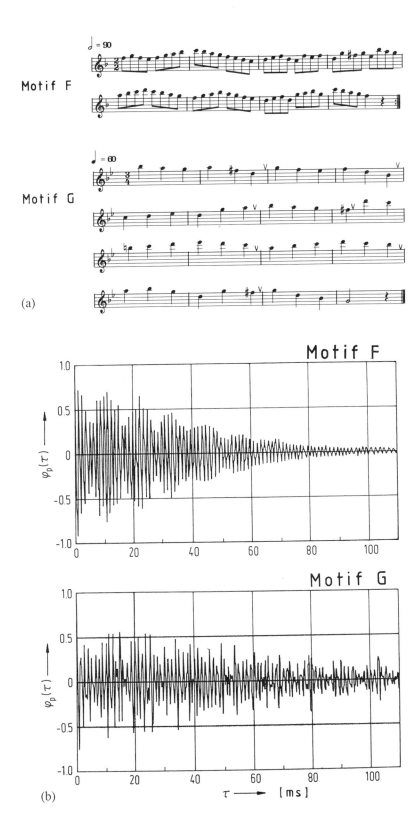

Motif F

Motif G

(a)

Motif F

$\varphi_p(\tau)$

Motif G

$\varphi_p(\tau)$

τ [ms]

(b)

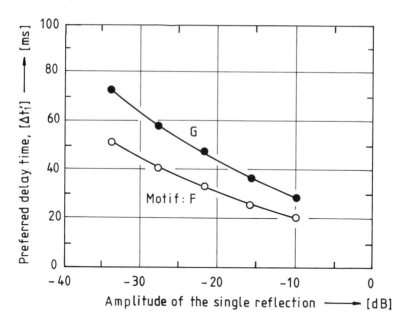

FIGURE 7.3. Most preferred delay time of a single reflection, $[\Delta t_1']_p$ as a function of the amplitude of reflection (Nakayama and Uehara, 1988; Nakayama, 1988). (o): Music Motif F; (●): Music Motif G.

amplitude of reflection is adjusted to be of equal loudness, then the most preferred delay time remains the same as in Figure 7.3.

Therefore, the most preferred delay time of the single reflection or the first reflection is well described by the duration τ_p' of the ACF, similar to Equation (4.4), which is defined by

$$[\Delta t_1']_p = \tau_p',$$

such that

$$|\phi_p(\tau)|_{\text{envelope}} \approx kA'^c \qquad \text{at} \quad \tau = \tau_p', \tag{7.1}$$

where the constants $k = 2/3, c = 1/4$, and A' is the total amplitude of reflections, being defined by $A' = 1$ relative to -10 dB of the direct sound as measured at the ear-canal entrance. This is due to the "missing reflection" phenomenon of performers (see Figure 6.11). Thus, for example, $A' = 1/4$ for -22 dB and $1/16$ for -34 dB.

If the envelope of the ACF is exponential, then Equation (4.5) may be applied, yielding

$$\tau_p' = (\log_{10} \frac{3}{2} - \frac{1}{4} \log_{10} A')\tau_e. \tag{7.2}$$

Figure 7.4 shows the relationship between the most-preferred delay time and the calculated value obtained by Equation (7.1) with the ACF-envelope and different amplitudes of reflection. The correlation coefficient between them is $0.99(p < 0.01)$. Table 7.1 demonstrates the procedure for obtaining the most-preferred delay time of the reflection approximately, by the use of Equation (7.2).

In order to obtain the preferred angles of incidence of a single reflection, paired-comparison tests were conducted with fixed loudness. The delay time of reflection was fixed to the most-preferred condition from Figure 7.3 for a constant amplitude of reflection. The resulting scale-values of preference as a function of the angle of incidence are shown in Figure 7.5. In this figure, the scale values for two music motifs were averaged, because no significant differences between them were achieved. It is found that the important directions of a reflection maximizing preference are the reflection from the rear ($\xi = 180°$) or from above ($\eta = 90°$). Thus, the location of reflection must be in the median plane (Nakayama, 1986). This result differs greatly from the condition of listeners who prefer the sound fields with a low value of the IACC. These results recommend that reflections from the rear wall and canopy or ceiling of the stage must be carefully designed for musicians, and controlled for music programs during rehearsals.

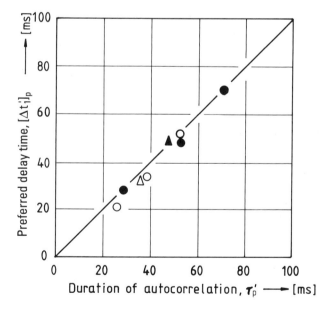

FIGURE 7.4. Relationship between the preferred delay of the single reflection and the duration of the ACF calculated by Equation (7.1) (Nakayama, 1988). Different symbols indicate the averaged values obtained by different motifs with different amplitudes of reflection ($A_1 = -10, -22,$ or -34 dB). ($\circ$): Music motif F; ($\bullet$): Music motif G; and ($\triangle$, $\blacktriangle$): results with two early reflections ($A_1 = A_2 = -22$ dB).

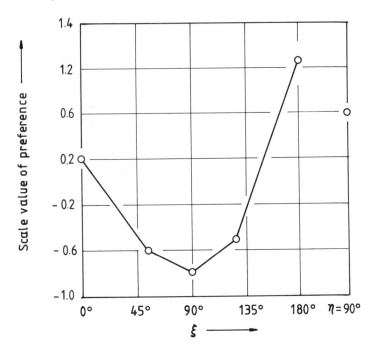

FIGURE 7.5. Scale values of preference as a function of the horizontal angle of incidence ξ of a single reflection to the subject (Nakayama and Uehara, 1988; Nakayama, 1988). The angle $\eta = 90°$ signifies that the single reflection arrives from above in the median plane.

TABLE 7.1. Judged and calculated preferred delay times of a single reflection for alto-recorder soloists. Judged values of $[\Delta t_1']_p$ are shown in Figure 7.3, and calculated values of $[\Delta t_1']_p$ are obtained by Equation (7.2) with values of τ_e.

A [dB]	A' [dB] (= A + 10)	A'	Judged $[\Delta t_1']_p$ [ms] Motif F	Motif G	Calculated $[\Delta t_1']_p$ [ms] Motif F	Motif G
−10	0	1.0	20.5	28.5	18.5	25.5
−16	−6	0.5	25.5	36.5	26.3	36.3
−22	−12	0.25	33.0	47.0	34.3	47.4
−28	−18	0.125	40.0	58.0	42.2	58.3
−34	−24	0.063	50.5	62.5	50.0	69.0

Next, it will be shown that this result holds for a sound field with two early reflections. A second reflection ($\xi = 0°, 54°, 90°$ or $126°$; $\eta = 90°$) was added to the single reflection at $\xi = 180°$ to determine the preferred temporal and spatial conditions of the second reflection. The first reflection was fixed at the preferred condition due to Equation (7.1) with the total amplitude of reflections A' (see Table 7.1). The scale values of preference obtained, for amplitudes of $A'_1 = -27$ dB and $A'_2 = -33$ dB ($A'_2 = A'_1 - 6$ [dB]), are shown in Figure 7.6 as a parameter of ($\Delta t'_2 - \Delta t'_1$).

The most-preferred delay time of the second reflection may be roughly found as

$$[\Delta t'_2]_p \approx 1.5[\Delta t'_1]_p. \tag{7.3}$$

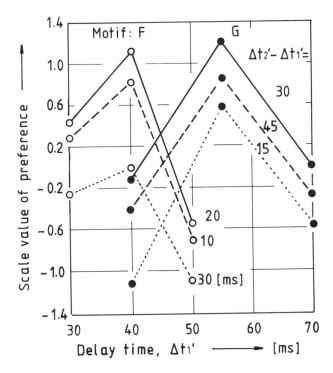

FIGURE 7.6. Scale-values of preference as a function of the delay time of the first reflection and as a parameter of the second reflection $\Delta t'_2 - \Delta t'_1$, for two different music motifs (Nakayama, 1988). The most-preferred delay times of the first reflection calculated by Equations (7.1a, b) are 40 ms and 55 ms for Music Motifs F and G, respectively ($A_1 = -27$ dB, $A_2 - A_1 = -6$ dB).

As shown in Figure 7.6, it is found that the preferred condition of the second reflection depends greatly on the source program played, but are hardly affected by the delay time of the first reflection $\Delta t'_1$ ($p < 0.01$) validating Equation (7.1). The preferred delay time between the direct sound and the second reflection $[\Delta t'_2 - \Delta t'_1]_p$ is about $0.5[\Delta t'_1]p$ shorter than that of the first reflection [Equation (7.3)], and may be associated with $0.8[\Delta t_1]_p$ for the condition of the listeners.

It is worth noting that the preferred incident angles of the second reflection are similar to the results of the first reflection, as shown in Figure 7.5.

7.2. Influence of the Music Program Selection on Performance

First of all, it is recommended that musicians select suitable music programs be fitting the concert hall with given temporal factors, the initial time-delay of the early reflection, and the reverberation time for listeners. Figure 7.7 shows the recommended ranges of reverberation time according to the range of τ_e of source signals for several music programs. It is quite natural that composers write music that images the acoustic condition of a certain room, with temporal factors represented by "reverberance." For example, music tempo and melody are slow enough for a pipe organ in a large cathedral with a long reverberation time. Chamber music

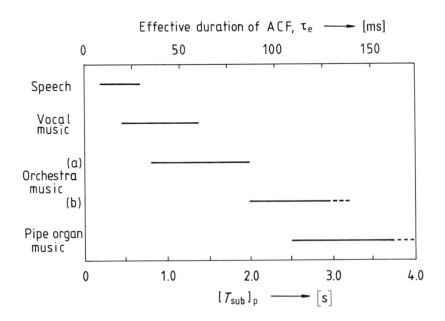

FIGURE 7.7. Estimated ranges of the effective duration of the ACF of sound sources, and the preferred reverberation times for listeners.

is performed in a small concert hall with short reverberation time. It is an interesting fact that fast high tempo music of Mozart was to be performed in the guest rooms of Courts with small audiences.

On the other hand, for music performers, the initial time-delay gap between the direct sound and the first reflection is controlled by adjusting the height of the canopies above the stage (Section 8.4), according to the ACF duration of music sources as discussed above. This kind of control may be exercised by a "sound coordinator," in the rehearsal prior to the performance.

Considering the fact that, for listeners, reflections from side walls including those on the stage are effective in decreasing the IACC, and that reflections from above the stage or from the rear wall of the stage are very important for performers, these conditions are typically realized at the same time without any serious contradictions.

7.3. Selection of the Performing Position for Maximizing Listener's Preference

The performing position that minimizes the IACC of the sound field for listeners' seats is demonstrated here by an example (Mouri, Mori, and Ando, unpublished). Scale values of preference are calculated with Music motif B (Arnold; $\tau_e = 35$ ms) at 112 listeners' positions in a Békésy Courtyard (Békésy, 1934). For simplicity, the directivity of the sound source is assumed to be uniform in this calculation. The height of the sound source is 80 cm from ground level, and the height of the listeners' ears is 110 cm. The reverberation time is 1.5 s.

The contour lines of equal average value of the IACC for 112 listeners' position, calculated to find the optimum performing positions, are plotted in Figure 7.8. The effective positions for performance may be found in the area minimizing the IACC for all of the listeners' positions. This indicates the importance of the side walls (near the performing position) in decreasing the IACC in the audience area. The most effective performing position is indicated by the symbol $[S]_p$ in Figure 7.9. When the sound source is located at $[S]_p$, the orthogonal factors (except for the reverberation time and related scale values of the subjective preference at each listener's position) are shown in Figure 7.9. The reverberation time used here is the one measured by Békésy (1.5 s). The preferred seating positions are found in the area centered on the scale values of -0.65. This reveals a good sound field in the courtyard without reflection from above. Békésy (1934, 1967) mentioned at that time (and nearly all listeners and musicians agreed), that the musical quality of the sound field in the courtyard was much better than in the concert hall where the orchestra usually played.

Bosse (1997) emphasized the importance of blending music performance and the concert hall as the heart of music. Professional musicians may change the style of performance during rehearsal, blending the music and sound field in a concert hall in the manner discussed in Section 3.2. For a given concert hall, after

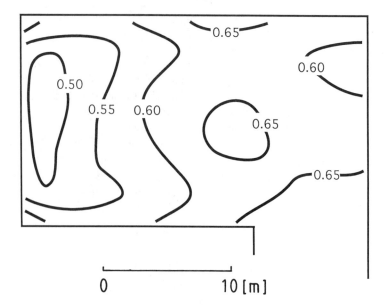

FIGURE 7.8. Contour lines of equal scale values of subjective preference calculated at each performing location averaged for 30 listening positions. The most effective location of music performance is found centered on $[S]_p$.

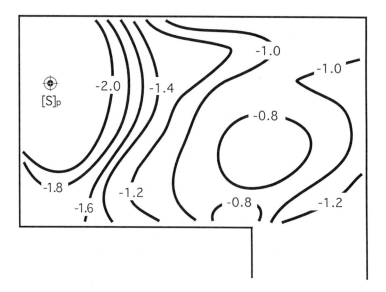

FIGURE 7.9. Contour lines of equal scale values of preference of listeners, when the performing position is located at the most effective source location $[S]_p$.

construction, musicians will learn what kind of music fit with the sound field, and how to control the style of performance of the music to be played. The acoustic factors in a concert hall are not significantly changed by aging over a long-time period.

8

Passive Control of Sound Field by Design

The acoustic design of each part of a room, improving the quality of the sound field at the listener's position, is demonstrated here, based mainly upon the spatial criteria.

8.1. Control of the IACC by Side Walls

The most effective and widely accepted factor in subjective preference judgment is the IACC. The shape of the concert hall can be best designed by minimizing the IACC, and thus maximizing the subjective preference, of listeners at each seat for any kind of sound source. The most remarkable fact is that the average values of interaural cross-correlation $\Phi_{lr}(0)$ for five music signals (Motifs A–E) indicate a minima at an incident angle of about $\xi = 55°$ from the median plane to the listeners, as shown in Figure 8.1. Behavior of the values of $\Phi_{ll}(0)$ and $\Phi_{rr}(0)$, which correspond to the average energies of the sound at the left and right ears, respectively, indicate a maximum difference at the angle centered on $\xi = 55°$.

In order to judge how a fundamental space may be formed, the IACC is calculated in the audience area by using the image method of a concert hall with a size of width W, length L, and height H, as shown in Figure 8.2. The floor area was fixed, and the stage area was fixed at one-third of the seat area. As discussed in Section 8.5, to minimize the attenuation of direct sound, the seating floor was inclined by 12° from the horizontal plane. For simplicity, the sound source was placed at the center of the stage, 1.5 m above the floor, and 20% of the distance from the front side of the stage. The absorption coefficient of the seating area was assumed to be 0.65. The simulation calculation was performed at 80–100 seat positions with the receiving point placed at the height of the listener's ear, 1.1 m.

The contour lines of equal IACC values (using Music Motif B) for each room shape are shown in Figure 8.3. It is well known that the IACC is increased near the sound source due to the strong direct sound. By narrowing the hall width W, the IACC is decreased, and the seat area for IACC < 0.5 is increased. This indicates the importance of sound reflection from the side walls. In order to control the IACC

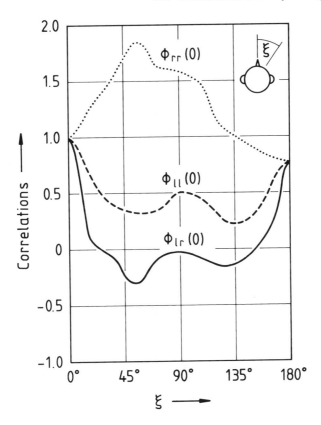

FIGURE 8.1. Average values [for five Music Motifs (A through E)] of the interaural cross-correlation, and the autocorrelation functions ($\tau = 0$) at the two ears for a single sound arriving from the horizontal angle ξ, which are needed for the calculation of the IACC($\tau = 0$). Values of the autocorrelation functions, $\Phi_{ll}(0)$ and $\Phi_{rr}(0)$, correspond to sound energies at each ears.

in the neighborhood of the source location, the angles of the side walls on the stage must be carefully designed (Section 10.1).

8.2. Control of the IACC by Ceilings

The effects of the variation of the ceiling surface angle, as shown in Figures 8.4 and 8.5, were examined. Figure 8.4 shows the Auditorium at Kobe University and Figure 8.5(a) Boston Symphony Hall with a height $H = 18$ m and Figure 8.6(b) shows the Boston Symphony Hall when the height H is adjusted to 9 m.

Similar to the above, the sound source was placed at the center of the stage and the subjective preference for the IACC values at specified seats were calculated.

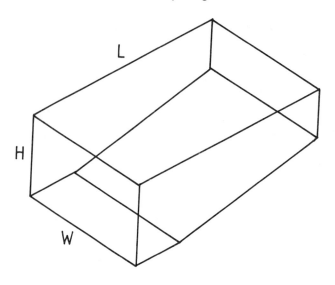

FIGURE 8.2. Simple geometry of a simulated concert hall.

The results of the scale values of the subjective preference S_4 due to the IACC (described in Section 4.4) are shown in Figure 8.6. When the ceiling angle is $\Psi = 30°$, the sound field of the auditorium at Kobe University is much improved up to the quality level in the Boston Symphony Hall. Even if the height of the Boston Symphony Hall is adjusted to be much lower than the existing level, the results are similar to those of the existing hall with $H = 18$ m. The most important fact is that, when the ceiling shape is flat, then the IACC is maximized and the scale value of preference is accordingly minimized.

8.3. Influence of Diffusers on Walls and Ceilings

8.3.1. Number Theory for Diffusers to Avoid the Image Shift

In the fundamental schemes of the concert hall described in the previous sections, Schroeder's diffusers may be added to the center part of the ceiling to avoid a strong reflection from the median plane. The diffuser design is illustrated in Figure 8.7. The design-frequency range of this diffuser is 250–1750 Hz (Schroeder, 1979). The reflection patterns for the diffuser are demonstrated in Figure 8.8. At the present stage of scientific knowledge, the calculation of the IACC at each seat in a concert hall with the Schroeder diffusers is difficult. But, it is interesting that the measured IACCs are smaller at the center parts of a hall than those calculated

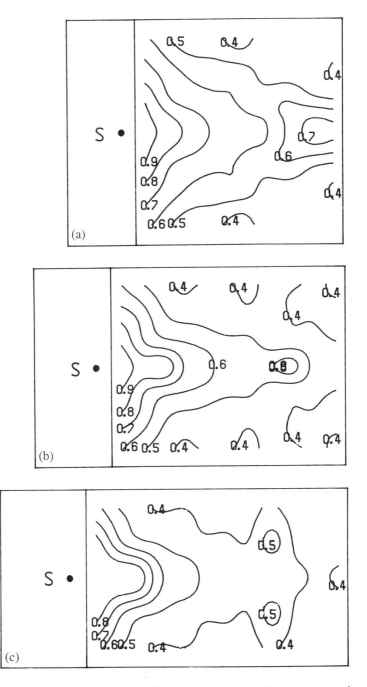

FIGURE 8.3. Contour lines of equal IACC values calculated with the geometry shown in Figure 8.2. (a) $L/W = 1.2$, $H/W = 0.6$; (b) $L/W = 1.6$, $H/W = 0.6$; and (c) $L/W = 2.0$, $H/W = 0.6$.

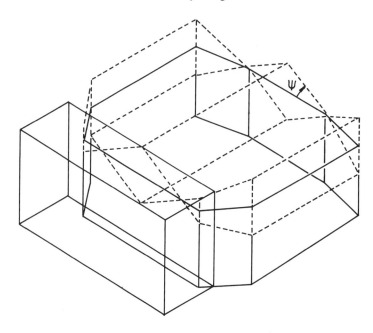

FIGURE 8.4. Auditorium at Kobe University with a ceiling angle Ψ varied in simulation, improving the sound-field quality with respect to the spatial factor, IACC.

without diffusers, showing that the diffusers close to the stage may be effective in reducing IACC as well (Ando et al., 1992).

The Schroeder's diffuser designed for the high-frequency range above 2 kHz may also be applied to side walls to avoid the image shift of the sound source for listeners. A deformed shape of this, excluding wells, was applied on the tilt-side walls in the Kirishima International Concert Hall (Miyama Conseru) as described in Section 10.1.

Since a significant amount of absorption due to the wells is found (Fujiwara, 1997; Onitzuka and Kawakami, 1997), a surface structure without wells will be discussed below.

8.3.2. Scattered Reflection by Uneven Surfaces

Other possible diffusers that avoid the image shift are realized by means of obstacles such as triangular forms and circular arcs on the reflecting walls. Calculated reflection patterns are shown in Figure 8.9. In this calculation, the diffusers are assumed as an infinite periodic structure, so that the reflection patterns are discrete as indicated by the arrows in these figures (Masuda and Fujiwara, 1997). When the dimension of the periodical surface is finite (5 m), then a continuous reflection pattern may be obtained instead of a discrete pattern (Figure 8.10). When the

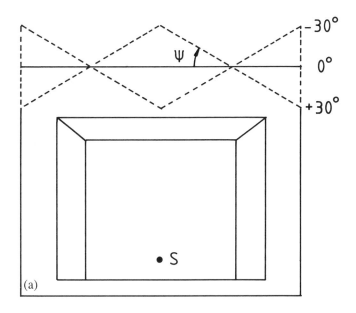

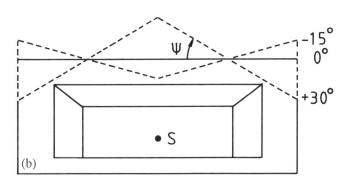

FIGURE 8.5. Boston Symphony Hall with a ceiling angle Ψ varied in simulation, improving the sound-field quality with respect to the spatial factor, IACC. (a) $H = 18$ m (original); and (b) $H = 9$ m (changed).

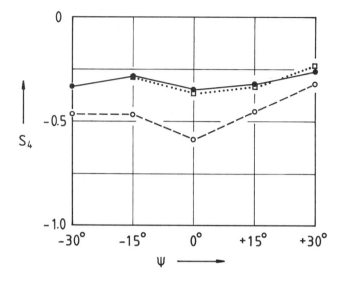

FIGURE 8.6. Scale-values of subjective preference S_4 in respect to the IACC (Music Motif B). (- - - - - -): Auditorium at Kobe University; (———): Boston Symphony Hall, $H = 18$ m (original); and (. . . . . .): Boston Symphony Hall, $H = 9$ m (simulated).

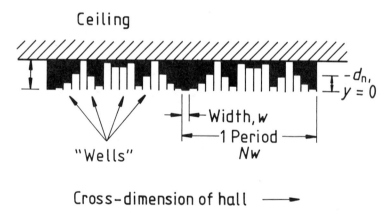

FIGURE 8.7. The two-dimensional Schroeder's diffuser based on quadratic residues of $N = 17$ (Schroeder, 1979).

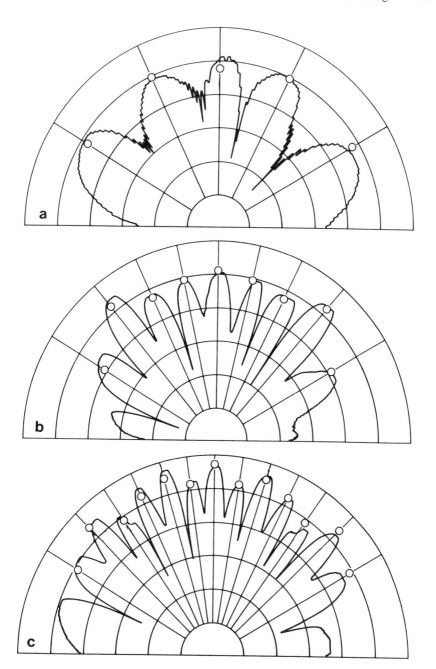

FIGURE 8.8. Measured and calculated reflection pattern for a scale model of the Schroeder's diffuser, $N = 17$ (Strube, 1980; 1981). (a) $\lambda_d/\lambda = 1.0$; (b) $\lambda_d/\lambda = 2.0$; and (c) $\lambda_d/\lambda = 3.0$.

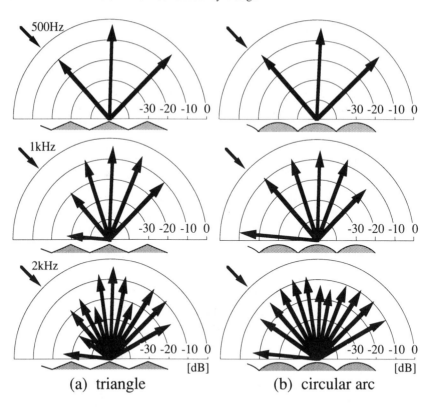

FIGURE 8.9. (a) Calculated directions of reflection for a two-dimensional periodic triangle shape; and (b) for the periodical circular-arc shape (Masuda and Fujiwara, 1997). The height of both shapes is 20 cm, and the period is 100 cm.

receiving point is close to the uneven surface, then a more diffusive pattern is observed, as shown in Figure 8.11. This is caused by a number of damped scattering waves that exist only near the uneven surfaces (Ando and Kato, 1976).

8.3.3. Fractal Geometry for Desired Sound Reflections

For a wider frequency range of reflections, fractal structures are proposed for the desired directions, as shown in Figure 8.12. In order to control the proper reflection angle to the listeners, according to Figure 6.3 for each frequency band, a fractal structure (Mandelbrot, 1982) may be applied as shown in Figure 8.12(b) and (c). Calculated results are shown in Figure 8.13 (Ando and Sakamoto, 1988; Dai and Ando, 1983). In order to minimize the IACC in each frequency band, three desired directions for the low-, middle- and high-frequency reflection may be realized by fractal geometry. For example, this kind of fractal geometry may be applied for

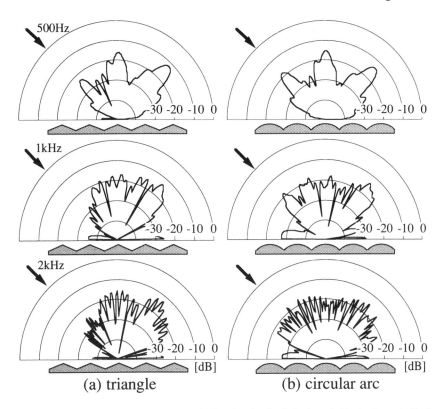

FIGURE 8.10. (a) Calculated directions of reflection for the finite length (5 m) of a periodical triangle shape; and (b) of a circular-arc shape (Masuda and Fujiwara, 1997). The receiving point is at the distance of 10 m from the center of the reflector. The amplitude of both shapes is 20 cm and the period is 100 cm.

side walls on the stage, obtaining more lateral reflections for the low-frequency range (below 500 Hz), and smaller angles to the median plane for listeners at higher frequencies. The complicated structures with the wells may be deformed more effectively to avoid excess absorption.

8.4. Reflectors near the Ceiling

8.4.1. Transfer Function for Reflection from Various Single Panel Shapes

Applying the Rubinowicz representation of the Kirchhoff diffraction integral, regarded as the mathematical formulation of Young's theory (Born and Wolf, 1970),

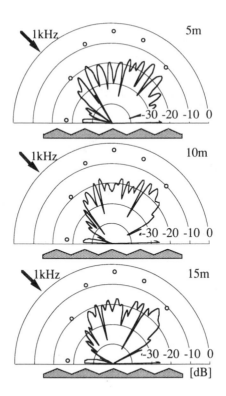

FIGURE 8.11. Calculated directions of reflection for the finite length (5 m) of the periodical triangle shape (Masuda and Fujiwara, 1997). The receiving point is varied from 5 m to 15 m from the center of the reflector. The amplitude of both shapes is 20 cm and the period is 100 cm. Circles denote the directional reflection levels shown in Figure 8.9(a).

which converts the surface integration into a line integration around the contour of a reflector, the transfer function for reflection of single panel reflectors is calculated (Nakajima, Ando, and Fujita, 1992). In order to obtain flat-frequency characteristics for reflection from a single plate, three types of single reflectors with a constant area of 4 m² are examined. The source point and receiving point are located at 14.14 m from the center of the reflectors. Results of the transfer function are shown in Figure 8.14 (a–c), as a parameter of the angle of sound incidence θ to the center of the plates. It is observed that the amplitude fluctuation of the transfer function increases with the number of sides of the polygon. Dips in the transfer function result from the simultaneous arrival of negative boundary waves at the receiving point from the edges, as indicated by the impulse responses shown on the right of Figure 8.14. Therefore, a triangular reflector is recommended to obtain flat-frequency characteristics. Further results show that the interior angle

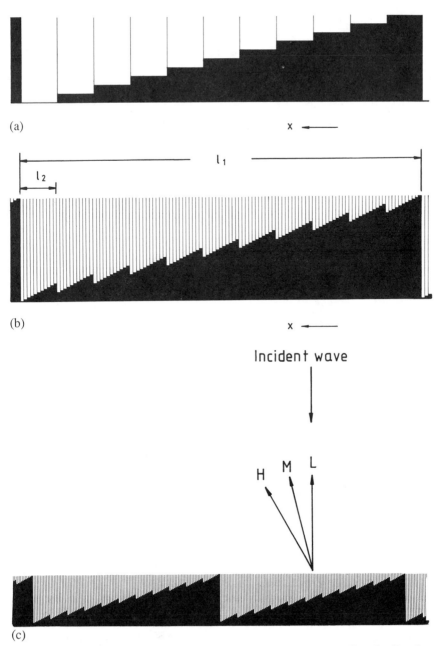

(a)

(b)

Incident wave

H   M   L

(c)

FIGURE 8.12. Formation of a fractal geometry for a reflector for controlling the direction of reflection for each frequency band. (a) Basic structure; (b) form of fractal geometry; and (c) desired directions, $\phi = 0°$, $-13.5°$, and $-29.8°$ of reflections for the low (L), middle (M), and high(H) frequency bands, respectively.

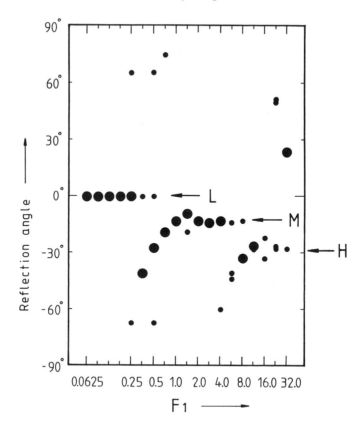

FIGURE 8.13. Calculated directions of reflection for the fractal geometry (Figure 8.12(c)), as a function of the normalized frequency F_1 for the three frequency bands, L, M, and H (Ando and Sakamoto, 1988). The normalized frequency is given by $F_1 = f/f_{d1}$, f_{d1} being the critical frequency between L and M to be designed.

of the isosceles triangular reflector is recommended to be in the range of 90° and 120°.

8.4.2. Transfer Function for Panel Arrays

In order to confirm the above results, the arrays shown in Figure 8.15 (a–c) composed of the three shapes of reflectors are examined. Each array has 35 panels, the total area of an array is 280 m², total panel area 140 m², so that the ratio of these areas is 50%. The transfer functions shown in these figures are calculated when

FIGURE 8.14. Calculated transfer function and the impulse response for reflection from reflector as a parameter of the angle of sound incident to the surface. (a) Triangle reflector; (b) square reflector; and (c) circular (decagon) reflector.

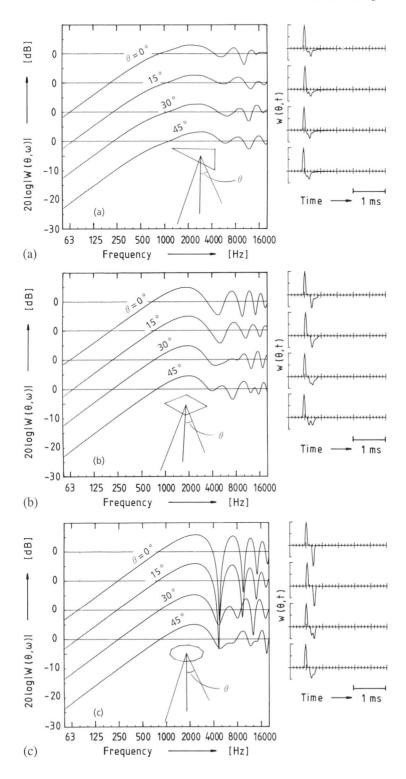

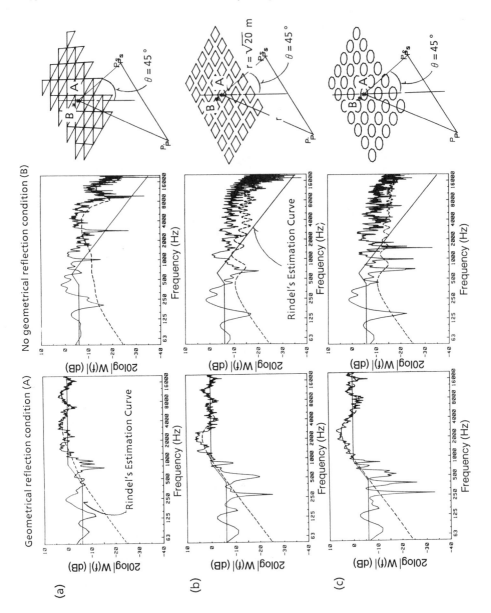

FIGURE 8.15. Calculated transfer function for the reflection of arrays from reflectors. (a) Array of triangle reflectors; (b) array of square reflectors; and (c) array of circular (decagon) reflectors.

the sound wave impinges the center of the array with the incident angle $\theta = 45°$. In Figure 8.15, the solid and dashed curves represent the calculated results for the panel arrays and for single panels, respectively.

Obviously, the dips in the transfer function of a triangular panel array in the low-frequency range (below 1000 Hz) are much smaller than the others. When a geometrical-ray reflection exists on a single panel, the transfer function in the high-frequency range is almost the same as that of the central single panel in the array. There are low-frequency components which do not exist for the reflection of a single panel. This phenomenon is caused by the diffraction effects of neighboring multiple panels as demonstrated in the next section.

For further information, the solid lines in Figure 8.15 indicate Rindel's estimation lines of the transfer function for a rectangular panel array (Rindel, 1986). The amplitude of transfer functions for panel arrays are in close agreement with Rindel's estimation only in the case when the path of geometrical reflection exists on the center of the panel.

8.4.3. Lateral Reflection Components from Canopies

In the Boston Symphony Orchestra's Tanglewood Music Shed, the canopy, which consists of nonplanar triangular panels, plays an important role in the decrease of the IACC, because there are no side-wall reflections (because of the fan shape of the Shed). The sides of the triangular panels range from 2.5 m to 8.0 m, and the opening spaces are triangular and of the same dimensions as the panels. The canopy is suspended about 6.5 m above the audience floor and extends over the stage as well as the frontal part of the audience area. Figure 8.16(a) is a typical example of the transfer function calculated for the panel array composed of 13 nonplanar triangular panels located within the ellipse drawn in the figure. The related impulse response is shown at the bottom of the figure. Figures 8.16(b–g) show the transfer function for individual panels, if all the neighboring panels are removed. In these figures, 0 dB refers to the level of the direct sound from the source to a receiving point without any reflection. It is remarkable that relatively strong low-frequency components arrive from panels away from the median plane as shown in the transfer functions. These reflections are adequate to decrease the IACC for the audio-frequency range. And, due to the high-frequency components from the panel in line above, it helps to avoid the image shift of the sound source, keeping the maximum value of the interaural cross-correlation function at the time origin, $\tau_{IACC} = 0$. Also, the low-frequency components near 200 Hz may compensate for the large attenuation due to the interference between the direct sound and the reflection of the seat rows and the floor (see Section 8.5).

In another example of an existing hall designed by Nakajima, Ando, and Fujita (1992), triangular reflectors are installed above and near the stage. Triangular reflectors with an angle of about 120° show the effective reflections for a wide-frequency range. When such reflectors are installed above the left and right stage,

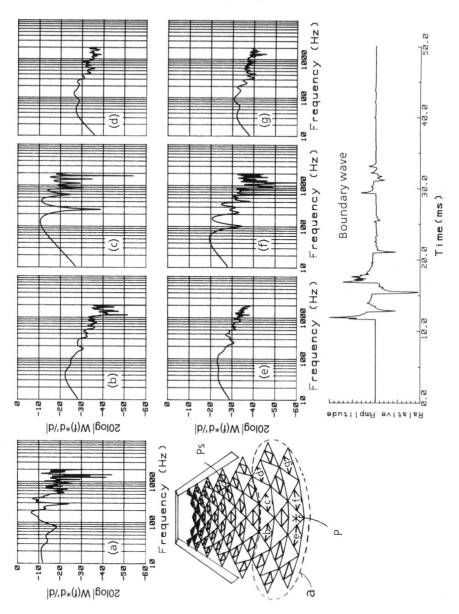

FIGURE 8.16. Calculated transfer function for reflection from a panel array composed of the 13 nonplanar triangular panels within an ellipse indicated by "a," which were installed in the Tanglewood Music Shed (a). The corresponding impulse response is indicated on the lower part of this figure. Calculated transfer function for reflection from each single nonplane triangular panel (b)–(g) within the ellipse without all neighboring panels (b)–(g). In the figures, 0 dB refers to the amplitude of the direct sound located at P_s (11.6 m, 6.5 m, 1.5 m) to a receiving point at P (24.0 m, 0.0 m, 1.5 m).

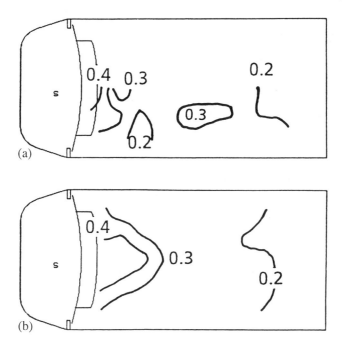

FIGURE 8.17. The IACC with music signal (Music Motif B) in an existing concert hall. (a) Measured values with the panel array composed of the seven nonplanar triangular panels installed above the stage; and (b) calculated values without the panel array.

then the lateral reflections in the low-frequency range may serve to decrease the IACC.

Results of the measured IACCs with the triangular reflectors above the stage are shown in Figure 8.17(a), and the calculated IACC without the reflectors are shown in Figure 8.17(b). When the reflectors are installed above the stage and/or near the stage, then the IACC values of seats close to the stage are decreased. According to the effective duration of the ACF of program sources, this kind of reflector above the stage is quite useful for musicians as well, supporting their performance by controlling the delay time of reflections from the height of canopy (Section 7.1).

8.5. Floor Structure and Seating

8.5.1. Low-Frequency Attenuation, and Effects of the Angle of Wave Incidence

In order to find the low-frequency attenuation and the effect of the angle of wave incidence θ, the sound transmissions over seat rows are calculated (Ando, Takaishi, and Tada, 1982). The angle of wave incidence is valid in the range of $\theta = 70°-89°$, and the acoustic admittance of the floor is kept constant at 0.2, which roughly corresponds to the absorption coefficient of the real floor with seats. Calculated results are shown in Figure 8.18 as a function of the frequency and as a parameter of the angle of wave incidence. The low-frequency attenuation appears around 100 Hz, which is independent of the angle of wave incidence. The sound pressure throughout the calculated frequency range decreases uniformly with an increasing angle of incidence. When the angle is kept smaller than 80°, then the excess attenuation remains less than 4 dB, except for the dip-frequency range.

It has been demonstrated that the maximum attenuation in the dip-frequency range diminishes with increasing absorption by the floor. The attenuation phenomenon is demonstrated in an existing hall by the measurement of the impulse response over the seat rows, as shown in Figure 8.19.

8.5.2. Effects of Under-Floor Cavities

The large attenuation occurring at the dip-frequency range can be reduced by making the floor sound absorbent. To accomplish this in the low-frequency range, the effects produced by slit resonators installed under the floor are examined.

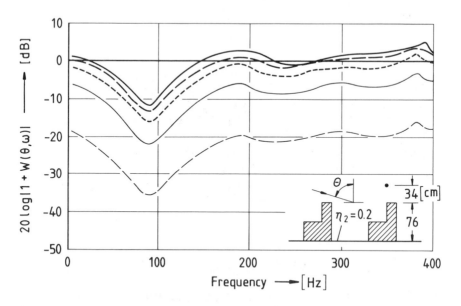

FIGURE 8.18. Calculated sound pressures over a seat rows with the angle of incidence q as a parameter (the period of the rows is 90 cm). The specific acoustic admittance of the floor surface is fixed at 0.2. (———):70°; (– – – – –):75°; (- - - - - -):80°; (———):85°; and (– – – – –):89°.

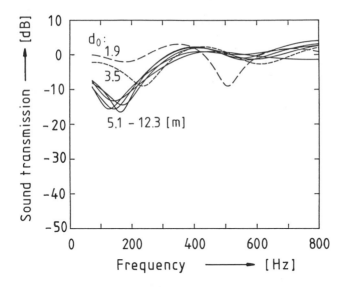

FIGURE 8.19. Measured sound pressure over the seat row in an existing auditorium, $\theta = 85°$. d_0 is a distance for the direct sound, from the source point to the listening position.

Figure 8.20 shows the sound pressure at ear level (without seats which are not effective in the low-frequency range) as a function of the frequency and as a parameter of the angle of incidence. There are no significant dips, indicating that sound absorption by the floor with a slit resonator is sufficient to remove the sound transmission dip. This characteristic is also demonstrated in calculating the seat rows.

A further practical alternative for improving the sound transmission over the seats and for decreasing the IACC is to design the space under the floor in a manner similar to that of the space above the ears. A typical example is shown in Figure 8.21, leaving an air gap, the bottom then has diffusing characteristics or similar shapes to the ceilings, as discussed in Section 8.2. This seating design has been adopted in the hall in Kobe, as described in Section 10.2.

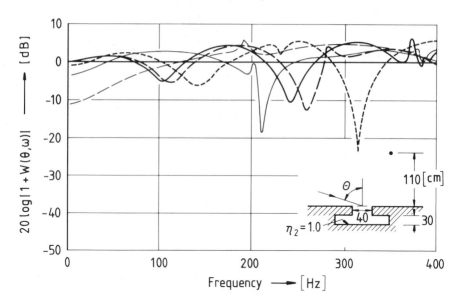

FIGURE 8.20. Calculated sound pressure over a slit resonator under the floor with the angle of incidence θ as a parameter without any seats (the period of the seats is 90 cm) (Ando, Takaishi, and Tada, 1983). (———):1°; (— — — —):21°; (- - - - - -):41°; (———):61°; and (— — — — —):70°.

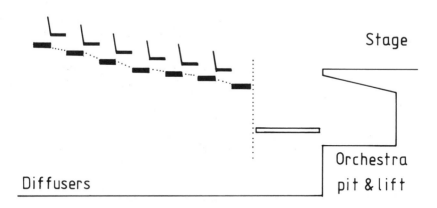

FIGURE 8.21. A schematic example of designing the structure under the floor [see also Figure 10.9(b)].

9

Individual Listener Subjective Preferences and Seat Selection

The minimum unit of audience is the individual. If each individual is satisfied by the environment, then the whole audience is satisfied. But the opposite is not true. Even if the condition of preference as a general standard is satisfied, as a general standard in the initial design, some listeners may not be satisfied. In this chapter, first of all, a method is discussed for determining individual subjective preference for the sound fields. Then, the results of individual differences are presented. As an example of this application, we now describe a seat selection system that enhances individual preference.

9.1. Individual Preference According to Orthogonal Factors

9.1.1. A Simple Method for Obtaining Individual Preference

Considering the fact that members of the audience, including children and older people, are quite diverse, a method for subjective judgments should be as simple as possible. For this purpose, the paired-comparison method is selected. Another method, for example, the method of the magnitude estimate, is too difficult for most people, except for skilled technicians in laboratory experiments. The paired-comparison method usually needs a number of judgments for a single pair. However, from a single observation datum for a set of sound fields, an approximate method is described to obtain the scale value of subjective preference. This method is based on the law of comparative judgment using the linear range of normal distribution between the probability and the scale value.

The probability that a sound field B is preferred to another sound field A is expressed by

$$P(B > A) = \frac{1}{\sigma_d \sqrt{2\pi}} \int_0^\infty \exp\left(-\frac{(X_d - \langle X_d \rangle)^2}{2\sigma_d}\right) dX_d$$

$$= \frac{1}{\sqrt{2\pi}} \int_{Z_{ab}}^\infty \exp\left(-\frac{y^2}{2}\right) dy,$$

(9.1)

where

$$Z_{ab} = -\frac{\langle X_d \rangle}{\sigma_d},$$

(9.2)

$\langle X_d \rangle$ is the average scale value between the sound fields A and B, $X_d = X_b - X_a$, if σ_d is being used as the unit for the scale value: $\sigma_d = 1$.

Thus,

$$P(B > A) = \text{erfc}[Z_{ab}],$$

(9.3)

$$Z_{ab} = \text{erfc}^{-1}[P(B > A)].$$

(9.4)

The first-order approximation of the Taylor series of Equation (9.4) is given by

$$Z_{ab} = \sqrt{2\pi}(P(B > A) - 1/2).$$

(9.5)

The linear range can be obtained for

$$0.05 \leq P(B > A) \leq 0.95.$$

Let us now consider a number of sound fields given by F $(i, j = 1, 2, \ldots, F)$, and suppose a single response from each pair for simplicity. Then the probability $P(B > A)$ in Equation (9.5) is replaced by (Ando and Singh, 1996)

$$P(i > j) = \frac{1}{F} \sum_{i=1}^F Y_i,$$

(9.6)

where $Y_i = 1$ responds to a preference of i over j, $Y_i = Y_j = 0.5$ ($i = j$), while $Y_i = 0$ corresponds to a preference of j over i. In order to improve the precision of the probability $P(i > j)$, a certain minimum number of sound fields within the linear range are needed to keep the accuracy, high when using Equation (9.6). This is performed by a preliminary investigation, avoiding an extreme sound field outside the linear range. In this manner, the scale value $S_i = Z_{ij}$ ($i = 1, 2, \ldots, F$) may be obtained approximately, when Z_{ij} with $P(j > i)$ is obtained by Equation (9.5).

Next, let us consider an error in a single observation. The poorness of fit for the model is defined by

$$\lambda = \frac{\sum_{(i,j)} |S_i - S_j|_{\text{Poor}}}{\sum_{(i,j)} |S_i - S_j|}, \qquad 0 \leq \lambda \leq 1,$$

(9.7)

where

$$|S_i - S_j|_{\text{Poor}} = S_j - S_i > 0 \qquad \text{if} \quad Y_i = 0,$$
$$= 0 \qquad\qquad \text{if} \quad Y_i = 1. \qquad (9.8)$$

Thus, in spite of j being preferred over i ($Y_i = 0$), it is possible that $S_j - S_i < 0$, and the amount $|S_i - S_j|_{\text{Poor}}$ is added, as in Equation (9.7). When i is preferred over j ($Y_i = 1$), it is natural that $S_i - S_j > 0$, and the amount is not added to the numerator. The value of λ corresponds to the average error of the scale value. This should be small enough, say, less than 10%.

Another observation is that, when the poorness number is K according to the condition expressed by (9.8), then the percentage of violations d is defined by

$$d = \frac{2K}{F(F-1)} \times 100. \qquad (9.9)$$

9.1.2. Examples of Individual Preference

Table 9.1 indicates typical examples of preference judgments with a single subject. The number of simulated sound fields is $F = 12$, with variations of both the listening level and the IACC. The value T_i is the aggregated preference scores of each sound field. For the scale values listed in Table 9.1, the number of violations $K = 6$ thus, $d = 9.1\%$, and $\lambda = 0.04$ (Singh and Ando, unpublished).

The results of scale values obtained as a function of the listening level and as a parameter of the IACC for a single subject with Music Motif A are shown in Figure 9.1. Almost parallel curves of values of the IACC are observed. This reveals that both the listening level and the IACC independently influence the subjective preference judgments. Hence, the scale values of preference may be described by each of the two factors, similar to the global preference with a number of subjects (Chapter 4). The most preferred listening level is always found to be close to 77 dBA for any value of the IACC. No interactive behavior may be found from the parallel curves due to change in the IACC, and similar curves relative to the listening level, in spite of the same right-hemispheric dominance (Sections 5.2 and 5.3). Smaller values of the IACC are always preferred regardless of the listening level. Thus, the scale values for the two factors are described, and are superposed in a manner similar to those described in Chapter 4. This kind of independence of the two factors was verified for all other 15 subjects participating.

In addition, such an independent nature may be found for the other two factors, both associated with the left hemisphere (Sections 5.2 and 5.3), the subsequent reverberation time and the scale of dimension SD of the hall or Δt_1 (= 22 SD), as demonstrated and summarized in Figure 9.2 for a single subject. The most-preferred reverberation time is always near 1.0 s for any value of SD, and the maximum preference is found near SD = 0.2 (Δt_1 = 4.4 ms; Music Motif B). Such independent behavior may well be achieved by means of the analysis of variance. Results are shown in Table 9.2 indicating that contributions of the factors are substantial enough to describe the total scale value without interference

TABLE 9.1. Example of scale values, S_i, estimated by aggregating the preference scores (0 or 1). The paired-comparison tests were conducted by changing both LL and IACC with Music Motif A (subject OS).

| Sound field | | | | | | | | | | | | | | | | |
LL[dB]	IACC		1	2	3	4	5	6	7	8	9	10	11	12	T	$P(i > j)$	S_i
83	0.98	1	0.5	0	0	0	0	0	1	0	0	0	0	0	1.5	0.13	−0.94
83	0.72	2	1	0.5	0	0	0	0	0	0	0	1	0	0	2.5	0.21	−0.73
83	0.39	3	1	1	0.5	1	0	0	0	0	0	1	0	0	4.5	0.38	−0.31
80	0.98	4	1	1	0	0.5	0	0	1	0	0	1	0	0	4.5	0.38	−0.31
80	0.72	5	1	1	1	1	0.5	0	1	0	0	1	1	1	8.5	0.71	0.52
80	0.39	6	1	1	1	1	1	0.5	1	0	1	1	1	1	10.5	0.88	0.94
77	0.98	7	0	1	1	0	0	0	0.5	1	0	1	1	0	5.5	0.49	−0.10
77	0.72	8	1	1	1	1	1	1	0	0.5	0	1	1	1	9.5	0.79	0.73
77	0.39	9	1	1	1	1	1	0	1	1	0.5	1	1	1	10.5	0.88	0.94
74	0.98	10	1	0	0	0	0	0	0	0	0	0.5	0	0	1.5	0.13	−0.94
74	0.72	11	1	1	1	1	0	0	0	0	0	1	0.5	0	5.5	0.49	−0.10
74	0.39	12	1	1	1	1	0	0	1	0	0	1	1	0.5	7.5	0.63	0.31

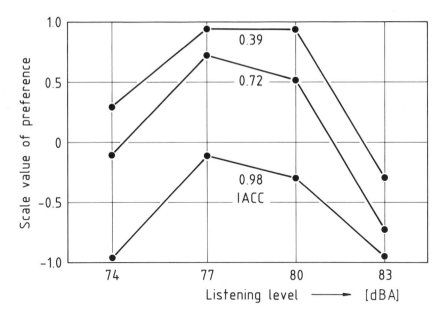

FIGURE 9.1. Scale values of preference for each sound field obtained by the paired-comparison test as a function of the LL and as a parameter of the IACC (subject OS, Music Motif A).

effects and errors. Under these conditions, the scale values of preference can be obtained by summation of the average values of each factor, because the scale value is linear. For all subjects tested, contributions of the factors are shown in Table 9.3. The amounts of the contribution of the factors differs by subject, but the total contribution is great enough to describe the scale value of preference, except for a few cases.

9.1.3. Individual Preference Description

Similar to the method described in Chapter 4 for a number of subjects, the scale values of subjective preference for each individual listener is also approximately expressed by (Ando and Singh, 1994)

$$S_i \approx -\alpha_i |x_i|^{3/2}, \qquad i = 1, 2, 3, 4. \qquad (9.10)$$

Therefore, the individual preference may be characterized by the coefficients $\alpha_i, i = 1, 2, 3, 4$, along with the positive and negative values of every x_i, and the most preferred values $[LL]_p$, $[\Delta t_1]_p$ and $[T_{sub}]_p$. For convenience, the positive and negative values of α_i, are averaged to obtain a single value.

The results of the preferred values of $[LL]_p$ and α_i, ($i = 1, 4$) are listed in Table 9.4, and the values of $[SD]_p$ and $[T_{sub}]_p$ are found in Table 9.5. Scale values of preference as a function of LL are shown in Figure 9.3 for individual subjects.

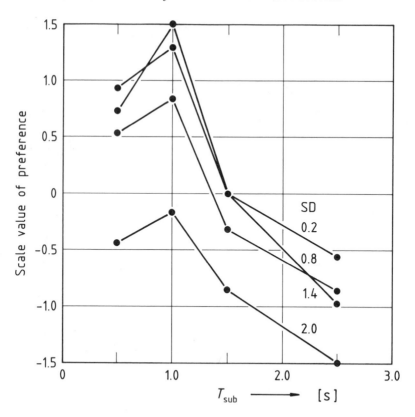

FIGURE 9.2. Scale values of preference for each sound field obtained by the paired-comparison test as a function of T_{sub} and as a parameter of the SD (subject OS, Music Motif B).

Obviously, the preferred listening levels greatly differ for each subject, exceeding the range tested, 74 dBA to 83 dBA. For example, subjects OK, AK, and YU preferred levels below 74 dBA, and subjects MA and HY preferred above 83 dBA.

The most remarkable results are shown in Figure 9.4, where all of the variation between individuals preferred a low value of the IACC without any exception and without regard for the two Music Motifs A and B used in the tests.

As shown in Figure 9.5(a) with Music Motif A, three subjects preferred delay around $[\Delta t_1]_p = 66$ ms ($[SD]_p = 3$), but others preferred less than 22 ms ($[SD]_p = 1$) and some more than $\Delta t_1 = 134$ ms. With Music Motif B as shown in Figure 9.5b, the most-preferred values are around $[\Delta t_1]_p = 18$ ms ($[SD]_p = 0.8$) or less than 4.4 ms ($[SD]_p = 0.2$).

As shown in Figure 9.6, great individual differences are also observed for subsequent reverberation time. With Music Motifs A, for example, several subjects preferred about 3.0 s, but others preferred more than 6.0 s or less than 1.5 s. With

TABLE 9.2.   Examples of the analysis of variance (subject OS).

Music	Factors	Degree of freedom	Sum of square	F-test	P	Contribution (%)
Motif A	Δt_1 (SD)	3	2.54	6.16	< 0.02	21.4
	T_{sub}	3	8.09	19.64	< 0.01	68.2
	Residual	9	1.24			
	LL	3	5.01	41.47	< 0.01	52.3
	IACC	2	4.33	53.71	< 0.01	45.2
	Residual	6	0.24			
Music B	Δt_1 (SD)	3	3.32	22.26	< 0.01	28.6
	T_{sub}	3	7.83	52.51	< 0.01	67.5
	Residual	9	0.45			
	LL	3	5.01	41.47	< 0.01	52.3
	IACC	2	4.33	53.71	< 0.01	45.2
	Residual	6	1.66			

Music Motif B, however, all the participating subjects preferred values near 1.0 s or less.

As described in the theory of subjective preference in Section 4.4, the most-preferred conditions for average of listeners (16 subjects) are: $[\Delta t_1]_p = 55$ ms (A = 3.7) and $[T_{sub}]_p = 2.9$ s for Music Motif A, $[\Delta t_1]_p = 14.5$ ms (A = 4.6) and $[T_{sub}]_p = 0.99$ s for Music Motif B. The preferred initial time-delay gaps also differ greatly among the individuals. But these most-preferred conditions are related well to the effective duration of the ACF of the source signal ($\tau_e = 127$ ms for Music Motif A; and $\tau_e = 43$ ms for Music Motif B) for each individual.

One of the reasons why such a great individual differences appeared in temporal factors and do not appear for the IACC, is that temporal factors have a greater influence on the individual brain during the time after birth in which the personality formed than on the spatial factor.

9.2. Effects of Lighting on Individual Subjective Preference

In order to gain knowledge of the interaction between lighting and an individual's subjective preference for the sound field, the lighting was controlled along with the listening level (LL), and the time delay of the single reflection (Δt_1) (Ando, Watanabe, and Yamamoto, 1990).

TABLE 9.3. Results of analyses of variance for all of subjects tested.

Subject	Music	Δt_1 (SD)	T_{sub}	Contribution (%) Total	LL	IACC	Total
OS	Motif A	21.4*	68.2**	89.6	52.3**	45.2**	97.5
	Motif B	28.6**	67.5**	96.1	52.3**	45.2**	97.5
MS	Motif A	57.7	14.2	71.9	—	—	
	Motif B	2.6	92.2**	94.8	49.0	24.4	73.4
HA	Motif A	9.8	63.7**	73.5	20.8*	71.2**	92.0
	Motif B	44.5**	51.5**	96.0	23.3**	72.2**	95.5
KO	Motif A	16.0	69.5**	85.5	25.1	52.2**	77.3
	Motif B	4.7*	91.7**	96.4	64.7**	21.3*	86.0
SK	Motif A	63.5**	12.3	75.8	53.6**	39.8**	93.4
	Motif B	4.0**	94.4**	98.4	29.7**	66.8**	96.5
MA	Motif A	4.6	79.5**	84.1	67.9**	26.8**	94.7
	Motif B	1.7	81.6**	83.3	36.4	29.2	65.6
YA	Motif A	—	—		19.5*	72.5**	92.0
	Motif B	5.2	84.3**	89.5	61.4**	34.0**	95.4
AK	Motif A	—	—		60.0**	32.1**	92.1
	Motif B	14.2*	79.0**	93.2	55.8**	33.4*	89.2
TA	Motif A	—	—		7.9	83.9**	91.8
	Motif B	4.1	88.8**	92.9	19.4	71.4**	90.8
TN	Motif A	58.9**	23.0*	81.9	47.0**	44.9**	91.9
	Motif B	16.8*	70.5**	87.3	9.8	64.9*	74.7
FU	Motif A	7.8	78.9**	86.7	42.1**	52.1**	94.2
	Motif B	24.6**	64.5**	89.1	58.8**	22.8	81.6
MI	Motif A	5.6	87.2**	92.8	45.8*	39.3*	85.1
	Motif B	4.2	92.3**	96.5	34.0*	55.5**	89.5
HY	Motif A	—	—		82.6**	10.7	93.3
	Motif B	19.5	60.8	80.3	86.5**	6.7	93.2
OK	Motif A	—	—		65.2**	25.2*	90.4
	Motif B	—	—		85.0**	8.8	93.8
MO	Motif A	—	—		29.2*	60.5**	89.7
	Motif B	—	—		8.3	60.4*	68.7
YU	Motif A	—	—		36.7*	54.7**	91.4
	Motif B	—	—		44.0*	43.0*	87.0

* $p < 0.05$
** $p < 0.01$

TABLE 9.4.   Preferred values of the LL and the weighting coefficients α_i ($i = 1$ and 4).

Subject	Motif A			Motif B		
	$[LL]_p$	α_1	α_4	$[LL]_p$	α_1	α_4
OS	78.0	0.15	1.12	77.3	0.06	1.26
MS	78.1	0.11	1.03	78.5	0.11	1.69
HA	76.5	0.07	1.31	80.1	0.07	1.70
KO	75.3	0.04	1.12	80.7	0.11	0.96
SK	78.3	0.13	1.22	79.8	0.10	1.52
MA	> 83.0	—	0.56	> 83.0	—	1.62
YA	80.0	0.09	1.17	> 83.0	—	0.99
AK	< 74.0	—	1.22	78.0	0.13	1.21
TA	78.5	0.06	1.40	76.9	0.08	1.76
TN	79.6	0.12	1.01	78.2	0.05	1.44
FU	77.2	0.08	1.22	80.4	0.08	0.61
MI	78.5	0.08	0.95	78.5	0.09	1.57
HY	> 83.0	—	0.29	> 83.0	—	0.47
OK	< 74.0	—	1.06	79.9	0.15	1.02
MO	76.8	0.07	1.21	77.0	0.06	1.39
YU	< 74.0	—	1.07	77.0	0.10	1.05

TABLE 9.5.   Preferred values of the factors SD (Δt_1) and T_{sub} and the weighting coefficients α_i ($i = 2$ and 3).

Subject	Motif A				Motif B			
	$[SD]_p$	$[T_{sub}]_p[s]$	α_2	α_3	$[SD]_p$	$[T_{sub}]_p[s]$	α_2	α_3
OS	< 1.0	< 1.50	—	—	0.74	0.82	3.38	11.46
MS	—	—	—	—	1.05	< 0.50	1.44	—
HA	2.56	3.43	2.46	5.03	< 0.20	0.90	—	7.45
KO	> 7.0	6.00	—	—	< 0.20	< 0.50	—	—
SK	< 1.0	3.42	—	6.98	< 0.20	< 0.50	—	—
MA	> 7.0	> 6.00	—	—	0.80	< 0.50	1.61	—
YA	—	—	—	—	0.88	0.82	2.20	7.35
AK	—	—	—	—	0.76	0.97	2.32	8.66
TA	—	—	—	—	< 0.20	0.91	—	8.88
TN	2.45	2.58	2.18	5.28	< 0.20	< 0.50	—	—
FU	3.72	< 1.50	4.12	—	0.76	1.12	1.91	4.22
MI	< 1.0	< 1.50	—	—	< 0.20	< 0.50	—	—
HY	—	—	—	—	< 0.20	0.83	—	3.30
TY	< 1.0	< 1.50	—	—	< 0.20	0.90	—	4.18

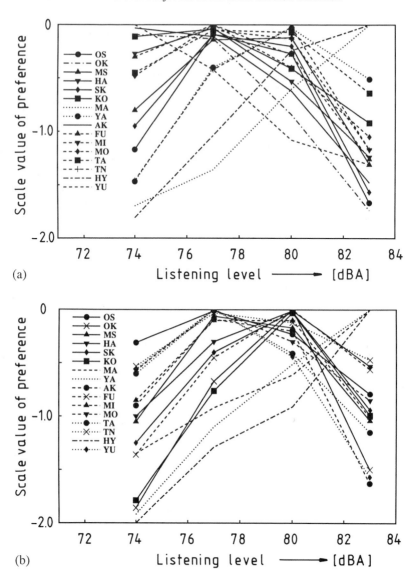

FIGURE 9.3. Individual differences of scale values of preference as a function of the LL. (a) Music Motif A; and (b) Music Motif B.

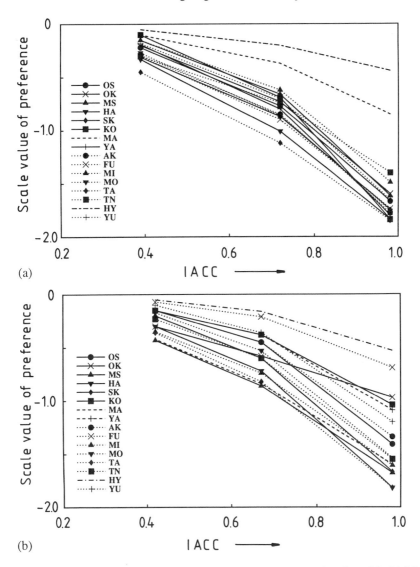

FIGURE 9.4. Individual differences of scale values of preference as a function of the IACC.
(a) Music Motif A; and (b) Music Motif B.

9.2.1. Effects on the Preferred Listening Level

As shown in Figure 9.7, a cotton screen transparent to both light and sound was used, resulting in a uniform illuminated environment. The intensity of the lighting varied over three levels, 2 lx, 35 lx, and 600 lx (brightness: 1.0 ± 0.2, 13.1 ± 1.5, and 183.7 ± 5.5, respectively). At the same time, the LL was controlled at five levels, -8 dB, -4 dB, 0 dB, $+4$ dB and $+8$ dB, in reference to the preferred LL

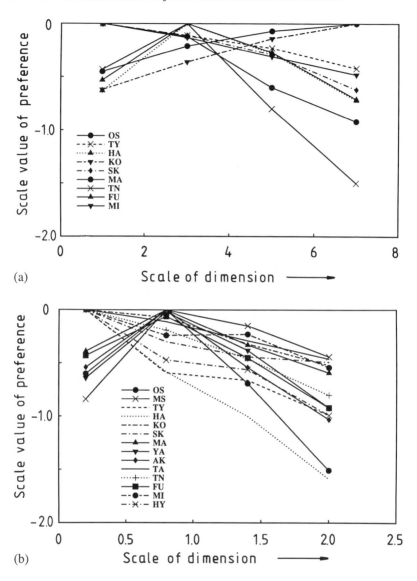

FIGURE 9.5. Individual differences of scale values of preference as a function of the ($\Delta t_1 \approx$ 22 SD). (a) Music Motif A; and (b) Music Motif B.

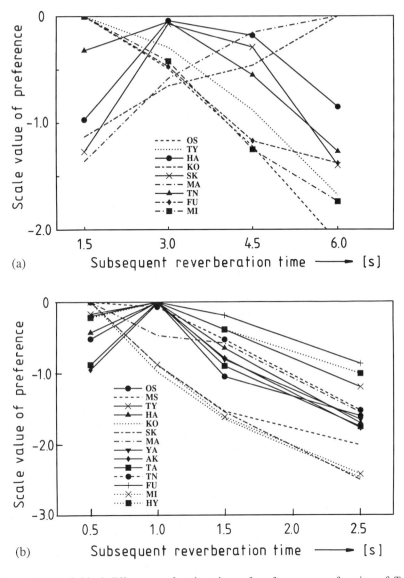

FIGURE 9.6. Individual differences of scale values of preference as a function of T_{sub}. (a) Music Motif A; and (b) Music Motif B.

FIGURE 9.7. Diagram of controlling the lighting level and physical factors of the sound field.

which was obtained by a method of adjustment as a preliminary test at 2 lx for each subject. This procedure is reasonable, because of the great individual differences in the preferred LL as mentioned in the previous section. The sound source was a speech signal, the reading of part of a poem (0.9 s). The subjects were seven students (six males and one female). Each subject was judged 16 times on asking up to 56 pairs extracted from 105 pairs in total $[n(n - 1)/2, n = 15]$.

The results of the scale value of the preference for the sound fields of each listener are shown in Figure 9.8. The significance level for the contributions of lighting to the preferred LL, obtained by the analysis of variance, are indicated in Table 9.6. The contributions of the lighting are less than 15%; however, four subjects A, B, D, and G, indicate significant results ($p < 0.05$). The most-preferred listening levels of subjects A, B, D, and E are somewhat shifted toward weaker sound-pressure levels at the high illumination of 600 lx. The preferred illumination during listening to the speech differed for each listener but, overall, about 35 lx appeared to be acceptable.

9.2.2. Effects on the Preferred Initial Time Delay

The subjects participating were eight male students (different from those in the above examination). The time delay of the single reflection gap was varied: $\Delta t_1 =$

TABLE 9.6.   Effects of lighting on the preferred LL.

Subject	Contribution of LL [%]	Contribution of lighting [%]	Preferred illuminance [1x]
A	85**	9*	2, 35
B	78**	13*	35
C	96**	0	600
D	86**	8*	—
E	97**	0	—
F	90**	4	—
G	84**	15**	35

* $p < 0.05$.
** $p < 0.01$.

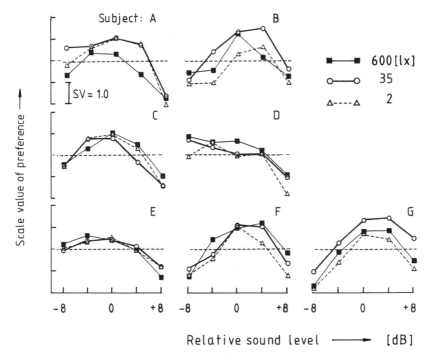

FIGURE 9.8. Scale values of preference of each subject as a function of the relative sound level (RL).

0 ms, 11 ms, 22 ms, 44 ms, and 88 ms ($A_0 = A_1 = 1$; $\tau_e = 12$ ms). Other experimental conditions were the same as those of the experiments above. The results of the scale value of preference are shown in Figure 9.9, and the contributions of the lighting to the preferred Δt_1 are indicated in Table 9.7. Despite large individual differences in preference judgments, for the sound field as mentioned above, almost all the subjects (except for subject O) indicated that the preferred values of Δt_1 are highly independent of the lighting levels.

Considering the fact that the change of lighting involves the right cerebral hemispheric dominance (Davis and Wada, 1974), no interference effects may be observed by a change in Δt_1 which involves the left-hemispheric dominance, as discussed in Section 5.2.2. On the contrary, much more interference effects are observed between the listening level and the lighting, because both are right-hemispheric dominances (see Section 5.2.2).

9.3. Inter-Individual Differences in Preference Judgments

As mentioned in Section 9.1.3, the individual difference is identified by the most-preferred value of each physical factor and the weighting coefficient α_i ($i = 1$, 2, 3, 4). In this section, the inter-individual difference is discussed with respect to the listening level ($i = 1$), such that

$$S_1 \approx -\alpha_1 |x_1|^{3/2}, \tag{9.11}$$

where $x_1 = LL - [LL]_p$ and α_1 is defined by a single value the positive and negative ranges of x_1. Here, fluctuations in terms of $[LL]_p$ and α_1 are examined for each test series for each subject (Sakai, Singh, and Ando, 1997). In this investigation, the Music Motif B was used as a source signal, and ten test series were conducted for each subject. Thirteen male students (21–24 years of age) participated as subjects. They had no previous experience in preference judgments. Five sound-pressure levels 66, 72, 78, 84, and 90 dBA were chosen to cover the preference range of the subjects participating.

In order to obtain values of $[LL]_p$ and α_1, a preference curve is drawn from the scale values obtained by the use of Equation (9.11) as shown in Figure 9.10. In this example, $[LL]_p = 75.0$ dBA and $\alpha_1 = 0.028$. The value of α_1 is obtained by the quasi-Newton method. All of the data were arranged in this way, and the resulting values of $[LL]_p$ and α_1 are shown in Figures 9.11 and 9.12, respectively, for each subject. Inter-individual differences of subjects C, E, F, J, and K were quite large, but not in the other cases. In order to discuss the reasons why such a large inter-individual difference was observed, scale values obtained in the ten test series for two extreme subjects K and G are plotted in Figure 9.13(a, b), in which values of $[LL]_p$ are shifted to 0 dB without any loss of information.

Obviously, if the value of α_1 is small, as it is for subject K, then the curves of scale values are rather flatter than those for subject G, and the most-preferred

TABLE 9.7.   Effects of lighting on the preferred Δt_1.

Subject	Contribution of Δt_1 [%]	Contribution of the lighting [%]	Preferred illuminance [lx]
H	97**	0	—
I	97**	0	—
J	92**	0	—
K	96**	0	—
L	71**	10	(2, 35)
M	63**	1	—
N	97**	0	—
O	94**	3*	35

* $p < 0.05$.
** $p < 0.01$.
( ) Only for the range of $\Delta t_1 > 11$ ms.

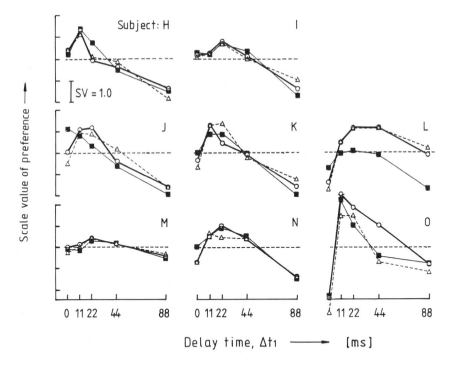

FIGURE 9.9. Scale values of preference of each subject as a function of Δt_1. Symbols are the same as those indicated in the upper right part of Figure 9.8.

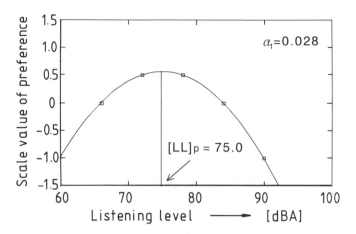

FIGURE 9.10. An example of the regression curve for scale values of subjective preference (subject G, Test series 2). $[LL]_p$ is found at 75.0 dBA.

listening levels are barely determined. On the other hand, for subject G, the value of α_1 is quite large and critical to determining the preferred level. Thus the inter-individual differences did not fluctuate depending on the number of test series. For all such critical subjects, the values of α_1 were greater than 0.02. The relationship between the range of fluctuation on the preferred listening level and the value of α_1 is demonstrated in Figure 9.14.

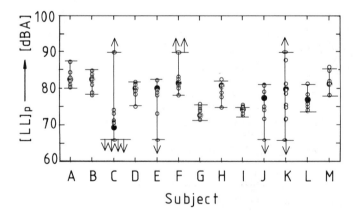

FIGURE 9.11. Results of inter-individual differences of $[LL]_p$ for each subject. Arrows indicate values of $[LL]_p$ that were out of the range 66 dbA to 90 dBA that was examined.

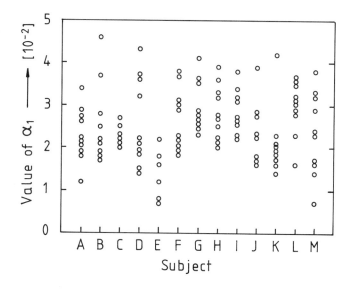

FIGURE 9.12. Values of weighting coefficient α_1 obtained from ten test series for each subject.

9.4. Seat Selection System for Individual Listening

9.4.1. Seat Selection System Enhancing Individual Preference

In order to maximize the individual subjective preference for each listener, a special facility of a seat selection system, testing his/her own subjective preference, was first arranged in use at the Kirishima International Concert Hall in 1994 (Sakurai, Korenaga, and Ando, 1997). The sound simulation is based on the system described in Section 3.5, with multiple loudspeakers as simplified in Figure 3.8. The system used allows for four listeners testing the subjective preference of the sound field at the same time. Since the four factors of the sound field influence the preference judgments almost independently, as was discussed in Section 9.1, each single factor is varied while the other three are fixed at nearly the most preferred conditions for a number of listeners. Results of the testing by acousticians who participated in the *First International Symposium on Music and Concert Hall Acoustics (MCHA95)*, which was held in Kirishima, in May 1995, are presented here.

9.4.2. Test Results of Individual Preference

The music source was orchestral, the "Water Music" by Handel; the effective duration of the autocorrelation function, τ_e, was 62 ms. The total number of listeners participating was 106. Typical examples of the test results as a function of each

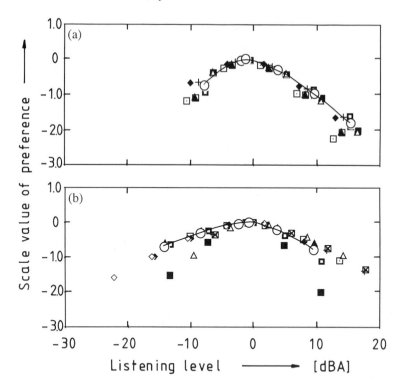

FIGURE 9.13. Examples of the scale values of preference obtained from ten test series. Different symbols correspond to results from the ten test series. (a) Subject G with a large value of α_1 (= 2.6×10^{-2}); and (b) subject K with a small value of α_1 (= 1.7×10^{-2}).

factor for the listener BL are shown in Figure 9.15. Scale values of the listener were close to the averages previously obtained: the most-preferred $[LL]_p$ is 83 dBA, $[\Delta t_1]_p$ is 26.8 ms (the preferred value calculated by Equation (4.5) was 24.8 ms, where $[\Delta t_1]_p \approx (1 - \log_{10} A)\tau_e$, A = 4.0), and the most-preferred reverberation time was 2.05 s (the preferred value calculated by Equation (4.7) is 1.43 s). Thus, the almost-center area of seats was preferred for listener BL as shown in Figure 9.16. Examples of the preferred value of each factor and the weighting coefficients for five listeners are listed in Table 9.8. With regard to the IACC, which is not listed in this table, it was the result for all listeners that the scale value of preference increased with the decreasing IACC value. Since listener KH preferred a very short delay time of the initial reflection, his preferred seats were located close to the boundary walls as shown in Figure 9.17. Listener KK indicated a preferred listening level exceeding 90 dBA (Table 9.8). For this listener, the frontal seating area close to the stage was preferable, as shown in Figure 9.18. For listener DP, whose preferred listening level was a rather weak 76.0 dBA, and preferred the initial delay time short; 15.0 ms, so that the preferred seats were in the rear part of

TABLE 9.8.   Examples of preferred value of each factor and the weighting coefficient represented by the individual difference in subjective preference.

Subject	Preferred LL[dBA]	Preferred $\Delta t_1[ms]$	Preferred $T_{sub}[s]$	α_1 $[10^{-2}]$	α_2	α_3	α_4
BL	83.0	26.8	2.05	6.0	1.86	1.46	1.96
KH	83.0	6.0	1.29	6.0	0.74	1.48	2.49
KK	> 90.0	21.0	1.84	1.0	1.39	0.83	2.02
DP	76.0	15.0	1.77	7.0	1.34	1.77	2.49
CA	83.0	> 100.0	1.27	1.0	0.30	1.45	2.84

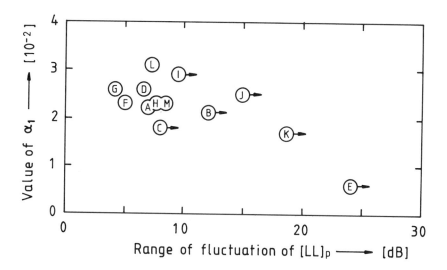

FIGURE 9.14. Relationship between the range of fluctuation of $[LL]_p$ and the weighting coefficient α_1.

the hall as shown in Figure 9.19. The preferred initial time-delay gap for listener AC exceeds 100 ms, but was not critical, as indicated by the value of $\alpha_2 = 0.30$. Thus any initial delay times are acceptable but the IACC is critical. Therefore, the preferred area of seats was located only in the center, as is shown in Figure 9.20.

9.4.3. Most-Preferred Conditions for Individuals

Cumulative frequencies of the preferred values with 106 listeners are shown in Figures 9.21 to 9.23 for three factors. As indicated in Figure 9.21, about 60% of

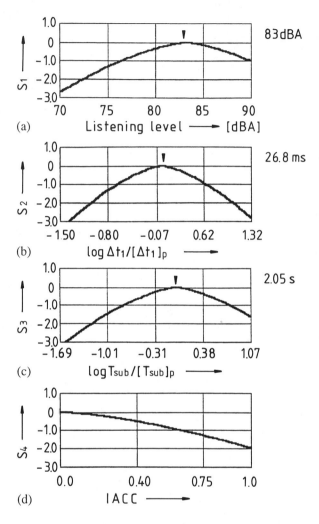

FIGURE 9.15. Scale values of preference obtained by tests for the four factors of the subject BL. (a) The most-preferred listening level is 83 dBA, the individual weighting coefficient in Equation (9.10): $\alpha_1 = 0.06$. (b) The preferred initial time-delay gap between the direct sound and the first reflection is 26.8 ms, the individual weighting coefficient in Equation (9.10): $\alpha_2 = 1.86$, where $[\Delta t_1]_p$ calculated by Equation (4.5) with $\tau_e = 62$ ms for the music used (A = 4.0) is 24.8 ms. (c) The preferred subsequent reverberation time is 2.05 s, the individual weighting coefficient in Equation (9.10): $\alpha_3 = 1.46$, where $[T_{sub}]_p$, calculated by Equation (4.11) with $\tau_e = 62$ ms for the music used, is 1.43 s; (d) Individual weighting coefficient in Equation (9.10): $\alpha_4 = 1.96$.

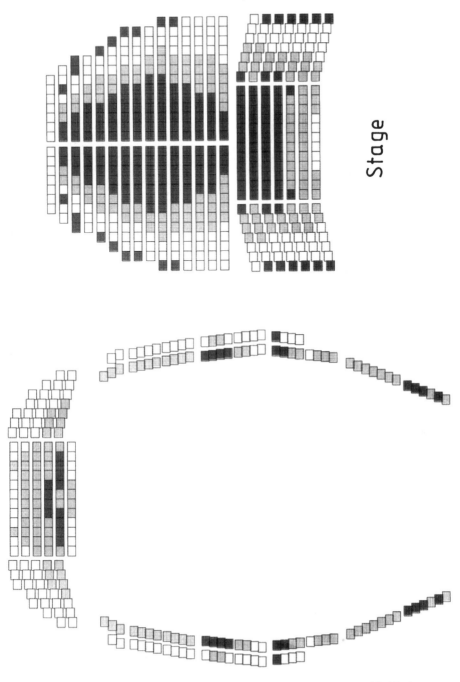

FIGURE 9.16. Preferred seat area calculated for subject BL. The seats are classified in three parts according to the scale values of preference calculated by the summation S_1 through S_4. Black seats indicate preferred areas about one-third of all seats in this concert hall for subject BL.

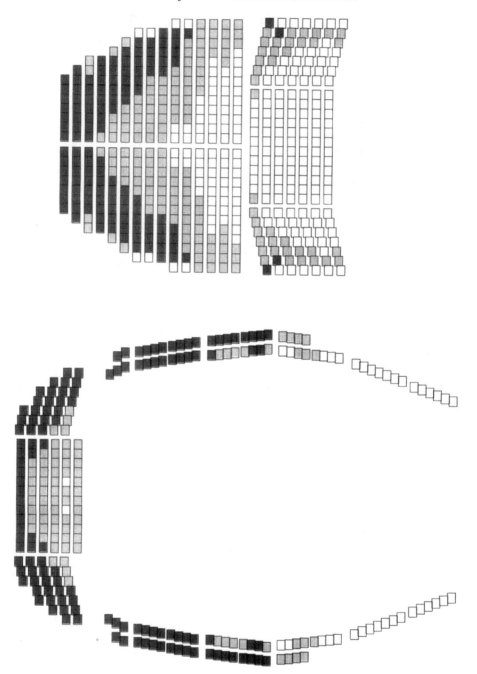

FIGURE 9.17. Preferred seat area calculated for subject KH.

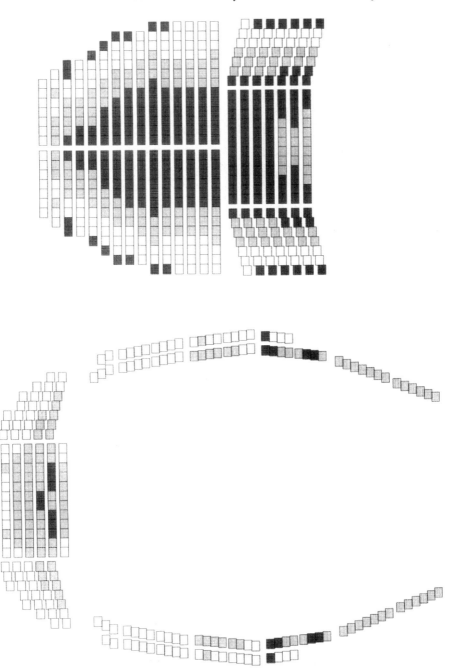

FIGURE 9.18. Preferred seat area calculated for subject KK.

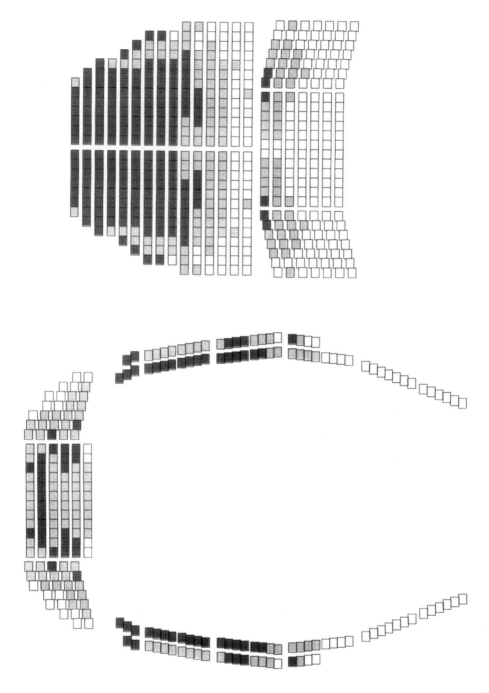

FIGURE 9.19. Preferred seat area calculated for subject DP.

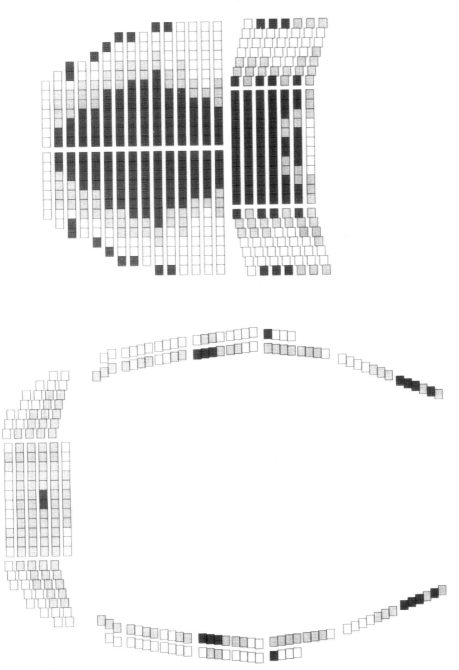

FIGURE 9.20. Preferred seat area calculated for subject CA.

listeners preferred the range 80 dBA to 84.9 dBA in listening to music, but some of the listeners indicated that the most-preferred LL was above 90 dBA, and the total range of the preferred LL was scattered, exceeding a 20 dBA range. As shown in Figure 9.22, about 45% of listeners preferred the initial delay times 20 ms to 39 ms which were around the calculated preference of 24.8 ms (Equation 4.5); some of the listeners indicated 0 ms to 9 ms and others more than 80 ms. With regard to the reverberation time, as shown in Figure 9.23, about 45% of listeners preferred 1.0 s to 1.9 s, which values are centered on the calculated preferred value of 1.43 s, but some listeners indicated preferences less than 0.9 s or more than 4.0 s.

Both the initial delay time and the subsequent reverberation time appear to be related to a kind of "liveness." Thus, it is assumed that there is a great interference effect between these factors for each individual. However, as shown in Figure 9.24, there is little correlation between the values of $[\Delta t_1]_p$ and $[T_{sub}]_p$ (correlation is 0.06). The same is true for the correlations between the values of $[T_{sub}]_p$ and $[LL]_p$; and for correlation between the values of $[LL]_p$ and $[\Delta t_1]_p$, a correlation of less than 0.11. Figure 9.25 shows the three-dimensional plots of the preferred values of $[LL]_p$, $[\Delta t_1]_p$, and $[T_{sub}]_p$. Looking at a continuous distribution in preferred values, no groupings of individuals can be seen to emerge from the data.

Cumulative frequencies of values of the α_i for four factors, $i = 1, 2, 3, 4$, are shown in Figures 9.26 to 9.29. These values signify weights for the four factors of each listener. For instance, when the value of α_1 is smaller than 0.02, then the listening level is insignificant for that listener, as discussed in the previous section.

Examples of the correlation between the values of α_1 and α_4 (both spatial factors), and the correlation between the values α_2 and α_3 (both temporal factors) are demonstrated in Figure 9.30, in which the values of correlation are less than 0.04. Since there are no correlations between the values of α_i and α_j ($i \neq j$), a listener indicating a relatively small value of α_i for one factor, will not always indicate a relatively small value for the other factor. Thus, a listener is critical about preferred conditions as a function of certain factors, while insensitive to other factors, resulting in a characteristic individual difference distinct from other listeners.

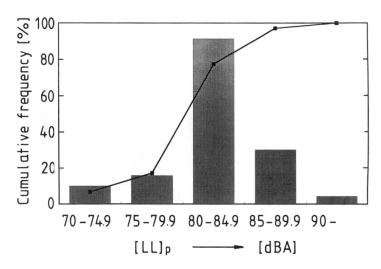

FIGURE 9.21. Cumulative frequency of the preferred listening level $[LL]_p$ (106 subjects). About 60% of subjects preferred the range of 80.0 dBA to 84.9 dBA.

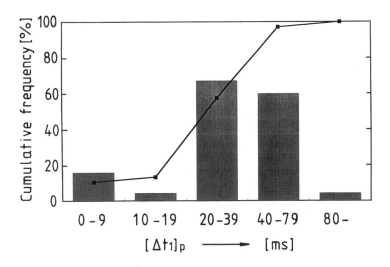

FIGURE 9.22. Cumulative frequency of the preferred initial time-delay gap between the direct sound and the first reflection $[\Delta t_1]_p$ (106 subjects). About 45% of subjects preferred in the range of 20 ms to 39 ms. The calculated value of $[\Delta t_1]_p$ by Equation (4.9) is 24.8 ms.

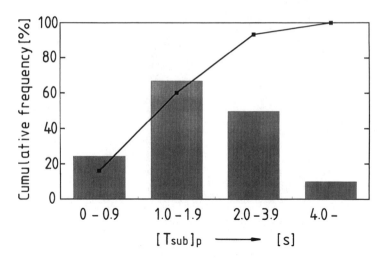

FIGURE 9.23. Cumulative frequency of the preferred subsequent reverberation time $[T_{sub}]_p$ (106 subjects). About 45% of subjects preferred the range of 1.0 s to 1.9 s. Calculated value of $[T_{sub}]_p$ by Equation (4.10) is 1.43 s.

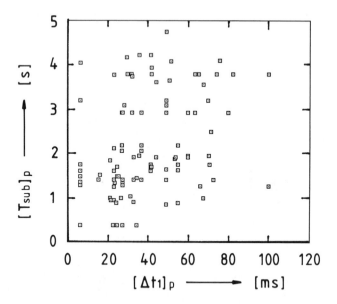

FIGURE 9.24. Relationship between preferred values of $[\Delta t_1]_p$ and $[T_{sub}]_p$ for each subject. No significant correlation between values was achieved.

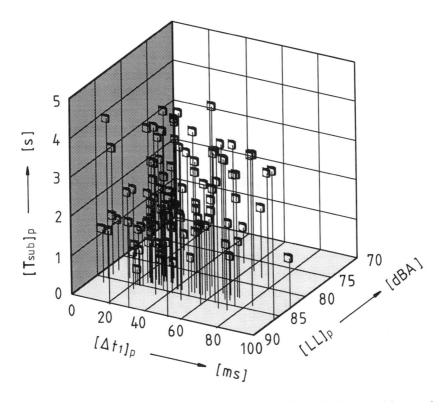

FIGURE 9.25. Three dimensional illustration of the preferred physical factors of the sound field for each subject. Preferred conditions are distributed in a certain range of each factor so that subjects could not be classified into any group.

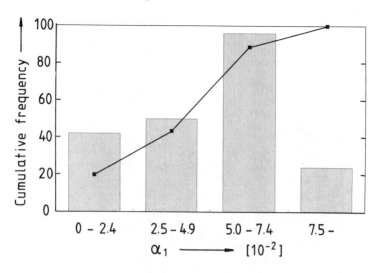

FIGURE 9.26. Cumulative frequency of the weighting coefficient α_1 of 106 subjects .

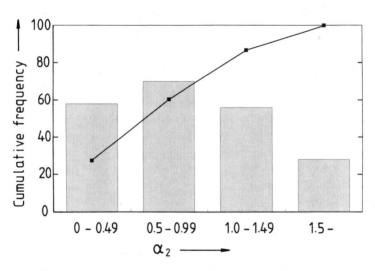

FIGURE 9.27. Cumulative frequency of the weighting coefficient α_2 of 106 subjects.

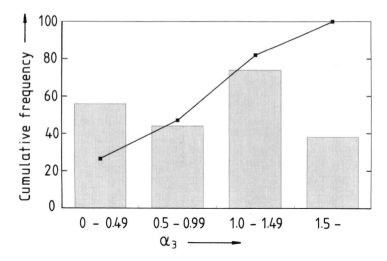

FIGURE 9.28. Cumulative frequency of the weighting coefficient α_3 of 106 subjects.

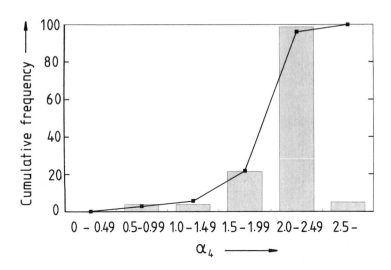

FIGURE 9.29. Cumulative frequency of the weighting coefficient α_4 of 106 subjects.

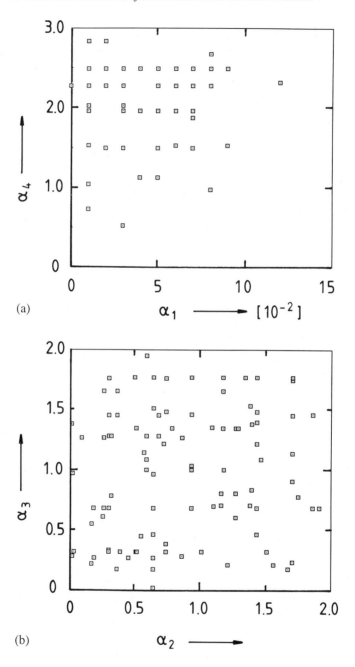

(a)

(b)

FIGURE 9.30. Relationship between values of α_1 and α_4 for each subject (a), and values of α_2 and α_3 for each subject (b). No significant correlations between values was achieved.

10

Case Studies of Acoustic Design

The purpose of this chapter is to demonstrate how the acoustic design of a concert hall and a multiple purpose auditorium is developed by means of case studies. As an example of outdoor spaces, the physical properties of a forest with multiple scattering phenomena are discussed in Chapter 11. There, it will be seen that the effects of scattering by a number of columns in an enclosure or by a number of trees in the outdoor space as elements of acoustic design may well be understood.

10.1. Concert Hall Design

The principle of the superposition of the scale value of subjective preference, with the optimal values of the four objective acoustics factors, can be applied to determine the total preference value at each seat. Comparison of the total preference values for different configurations of a concert hall allows us to choose the best design for a specific performance space, such as for a certain music program.

Procedures for designing sound fields in a concert hall are illustrated in Figure 10.1. Temporal factors and spatial factors are carefully designed, in order to satisfy both left and right human cerebral hemispheres for each listener, for the conductor, and for each musician on the stage.

(1) Temporal Factors for Listeners

First of all, the purpose of a concert hall under planning is determined by a classification of the music to be performed, with respect to a range of τ_e. The planning is associated with other facilities existing near the site, and with the location of the concert hall under design. If the space is designed for the performance of a pipe organ, the temporal factors Δt_1 and T_{sub} are determined by the range of τ_e which may be selected to be, say, centered about 200 ms ($T_{sub} \sim 4.6$ s, Section 4.3). When it is designed for the performance of chamber-music, the range of τ_e is selected near the value of 65 ms ($T_{sub} \sim 1.5$ s). The conductor or the sound coordinator selects suitable music motifs with a satisfactory range of τ_e of the ACF to achieve a music performance that blends the music and the sound field in the hall.

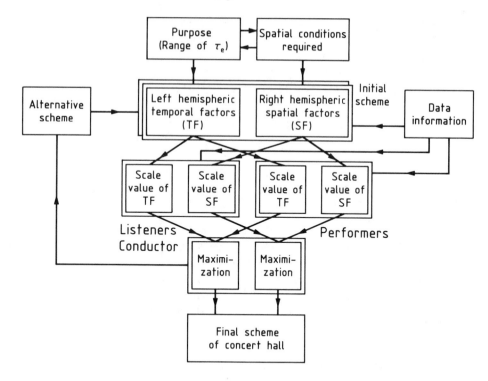

FIGURE 10.1. Acoustic-design procedure maximizing the scale value of both temporal factors (TF) and spatial factors (SF) for the sound field in a concert hall, enhancing the satisfaction for both human cerebral hemispheres. The specialization of the left hemisphere for temporal factors and the right hemisphere for spatial factors for each listener, conductor, and performer are taken into consideration.

The information for the ACF of music signals, and related subjective attributes are integrated. Moreover, in order to adjust the preferred initial time-delay gap for each music performance location, the position of each instrument is carefully placed on the stage. For instance, if the values of τ_e for violins is shorter than that of contrabasses with mainly the low-frequency ranges, the position of violins is shifted closer to the left wall on the stage and the position of contrabasses is shifted closer to the center as viewed from the audience.

(2) Spatial Factors for Listeners

As discussed in Chapter 8, the IACC should be kept as small as possible, maintaining $\tau_{IACC} = 0$. This is realized by suppressing the strong reflection from the ceiling, and by appropriate reflections from the side walls at particular angles. If the source signal contains mainly the frequency components around 1 kHz, the reflections from the side walls are adjusted to be centered at roughly 55° to each listener, measured from the median plane of the listener. Under actual hearing conditions in an existing hall, the perceived IACC depends on whether or not the amplitudes of reflection exceed the hearing threshold level, in addition to the physical value given by Equation (3.25). Thus, a more diffuse sound field may be perceived with increasing power of the sound source. For example, Keet (1968) reported the apparent source width (ASW) increases with an increase in the listening levels. While the source is weak enough, hearing only the direct sound, the actual IACC being processed in the auditory–brain system approaches unity, resulting in no diffuse sound impression. In general, small values of the IACC have to be realized by early strong reflections only. If the sound source is located on the center line on the stage, then coherent signals arrive at the same time from both the side walls. From this point of view, acoustical-asymmetric properties in shaping the hall may create further advantages.

(3) The Sound Field for Musicians

For music performers (alto-recorder), the temporal factors are much more critical than the spatial factors (Section 7.1). Since musicians perform over a sequence of time, reflections with a suitable delay time relative to the values of τ_e of the source signals is of particular importance (for cellists; Sato, Ota, and Ando, 1998). Without any spatial subjective diffuseness, the preferred directions of reflections are in the median plane of music performers, resulting in IACC $\approx$ 1.0. In order to satisfy these acoustic conditions, some design iterations are required, maximizing the scale values for both musicians and listeners, and leading to the final scheme of the concert hall (Section 7.2).

(4) The Sound Field for the Conductor

It is recommended that the sound field for the conductor should be designed as that of a "listener" with the appropriate reflections of the side walls on the stage (Meyer, 1995).

(5) Fusing Acoustic Design with Architecture

From the historical viewpoint, architects have been perhaps more concerned with spatial criteria from the visual standpoint and were less so with the temporal criteria for blending human life and the environment under design, while acousticians have mainly been concerned with the temporal criteria, represented by the reverberation time from the time of Sabine (1900 onward). There has existed no theory of design, by including the spatial criterion represented by the IACC, so that discussions by acousticians and architects were never on the same subject. As is described by the general theory of physical environments in Chapter 12, both temporal and spatial factors are deeply concerned with both acoustic design and architectural design.

As an initial design sketch of the Kirishima International Concert Hall, Kagoshima, Japan, a plan, shaped like a leaf (Figure 10.2) was presented at the first meeting for further discussion with the architect, Fumihiko Maki, with the explanation of temporal and spatial factors of the sound field (Ando, 1985). After some weeks, Maki presented a revised scheme of the concert hall as shown in

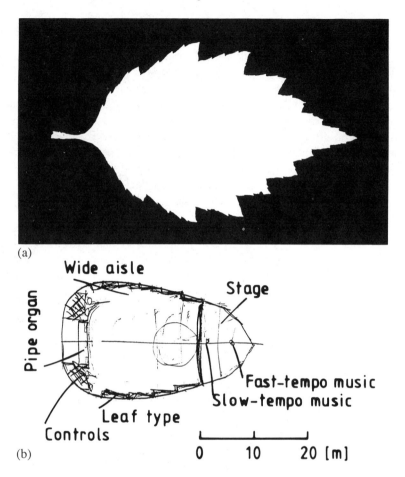

FIGURE 10.2. A leaf shape of the plan proposed for the Kirishima International Concert Hall, Kagoshima, Japan. (a) Original leaf shape (asymmetry); and (b) proposed asymmetrical shape for the plan. The sound field for seats in the circle are well designed for reflections from the walls on the stage and side walls.

Figure 10.3 (Maki, 1994(a, b); Maki, 1997; Ikeda, 1997). Without any change of plan and cross sections, the calculated results indicated excellent sound fields, as shown in Figures 10.4 and 10.5 (Nakajima and Ando, 1997).

The final architectural schemes, together with the special listening room for testing individual preference of the sound field and selecting the appropriate seats for maximizing individual preference of the sound field, are shown in Figure 10.6. In these figures, the concert courtyard, the small concert hall, several rehearsal rooms, and dressing rooms are also indicated. The Kirishima International Concert Hall under construction is shown in Photo 10.1, in which the leaf shape may be seen.

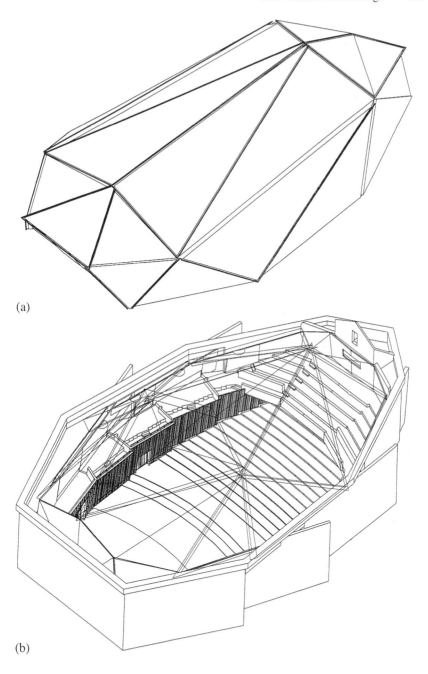

(a)

(b)

FIGURE 10.3. A scheme designed by the architect Fumihiko Maki (1994a, b).

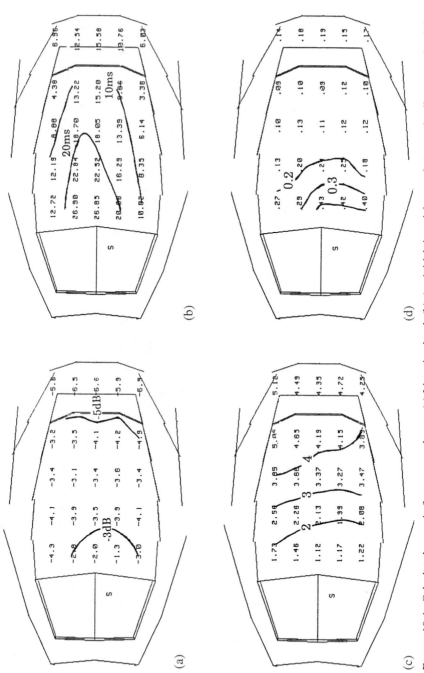

FIGURE 10.4. Calculated acoustic factors at each seat. (a) Listening level; (b) Δt_1, initial time-delay gap between the direct sound and the first reflection; (c) A-value, the total amplitude of reflections; and (d) IACC. The reverberation time designed was about 1.7 s for the 500 Hz band.

(6) Details of Acoustic Design

(a) For Listeners on the Main Floor

In order to obtain a small value of the IACC for most listeners, the ceilings consisted of a number of triangular plates with adjusted angles (Section 8.2), and the side walls were given a 10% tilt with respect to the main-audience floor (Section 8.1), as is shown in Figure 10.7 as well as in Figure 10.3. In addition, diffusing elements are designed on the side walls to avoid the image shift of sound sources on the stage caused by the strong reflection in the high-frequency range above 2 kHz. These diffusers on the side walls are designed by a deformation of the Schroeder's diffuser described in Section 8.3, taking the wells away, as shown by the detail of Figure 10.8(a).

(b) For Music Performers on the Stage

In order to provide reflections from places near the median plane of each of the performers on the stage, the back wall on the stage is carefully designed as shown at the lower left in Figure 10.7. The tilted back wall consists of six subwalls with angles adjusted to provide the appropriate reflections within the median planes of the performers. It is worth noticing that the tilted side walls on the stage provide good reflections to the audience sitting close to the stage, resulting in a decrease of the IACC. Also, the side wall may provide the effective reflection (arriving from the back) for a piano soloist.

(c) Stage Floor Structure

For the purpose of suppressing the normal-mode vibration (Morse, 1948) of the stage floor and an anomalous sound radiation from the stage floor during performances, the joists form triangles without any neighboring parallel structure, as shown in Figure 10.8(b). The thickness of the floor is designed to be thin, 27 mm, in order to radiate sound effectively through the vibration from instruments like the cello and contrabass. During rehearsal, music performers may control radiation power somewhat by adjusting their positions and/or by the use of a rubber pad between the floor and the instrument. There are, however, many problems to be solved for a more adequate design of the floor vibrations and the related sound radiation from the floor in connection with the direct sound radiation from the musical instrument itself. This hall was opened in July 1994 (Photo 10.2)

10.2. Multiple-Purpose Auditoria

As discussed in Chapter 9, all the listeners tested preferred a small value of the IACC as a spatial factor of the sound field; therefore, the IACC at each seat must be controlled by the spatial characteristics of the reflectors in the room. The main purpose of the auditorium to be designed is accommodated by the association of likely source signals with the values of the effective duration of the ACF. For example,

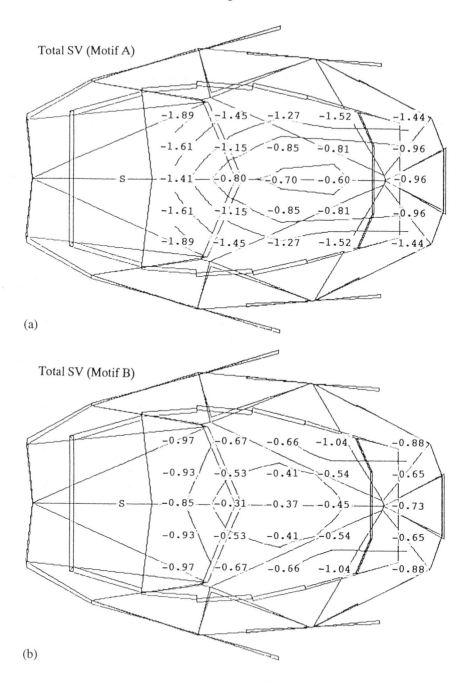

FIGURE 10.5A, B. Calculated total subjective scale values at each seat. (a) Performing position: Stage front with Music Motif A. (b) Performing position: Stage rear with Music Motif B. The performing position of the stage rear part near to the wall is recommended.

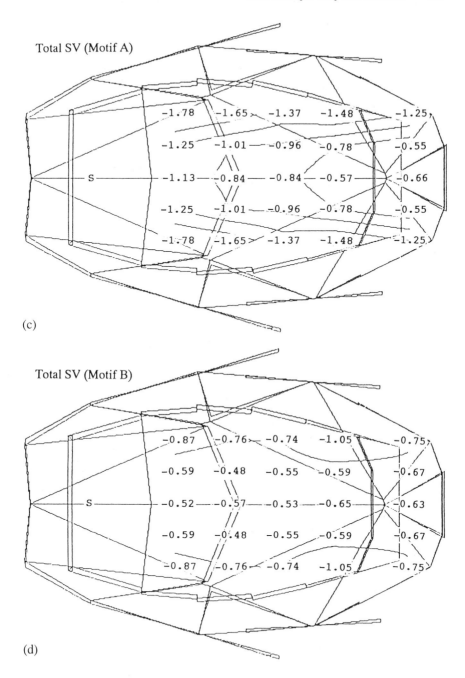

FIGURE 10.5c, d. Calculated total scale values at each seat. (c) Performing position: stage front with Music Motif A. (d) Performing position: stage rear with Music Motif B. The performing position of the stage rear part near to the wall is recommended.

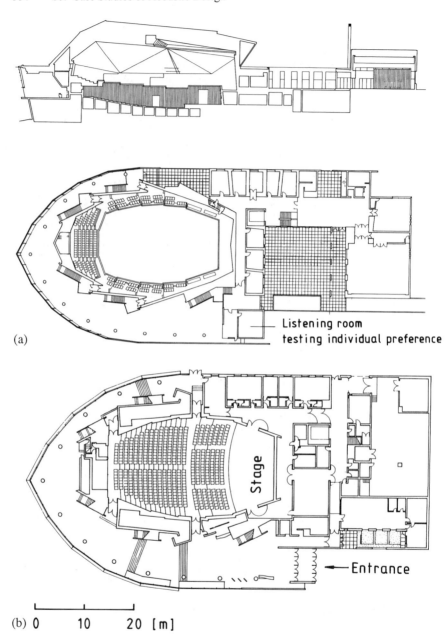

(a)

Listening room
testing individual preference

Stage

Entrance

(b) 0   10   20 [m]

FIGURE 10.6A, B. The final scheme of the Kirishima International Concert Hall, Kagoshima, Japan designed by the architect Maki (1994a, b). (a) Longitudinal section and plan of balcony level; (b) plan of audience level.

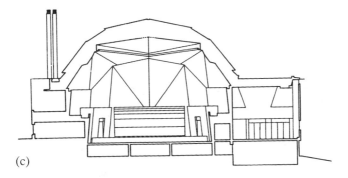

(c)

FIGURE 10.6c. The final scheme of the Kirishima International Concert Hall, Kagoshima, Japan designed by the architect Maki (1994a, b). (c) Cross-section.

PHOTO 10.1. The Kirishima International Concert Hall, Kagoshima, Japan, under construction. The leaf shape may be seen in the center part (photo by Yamamoto).

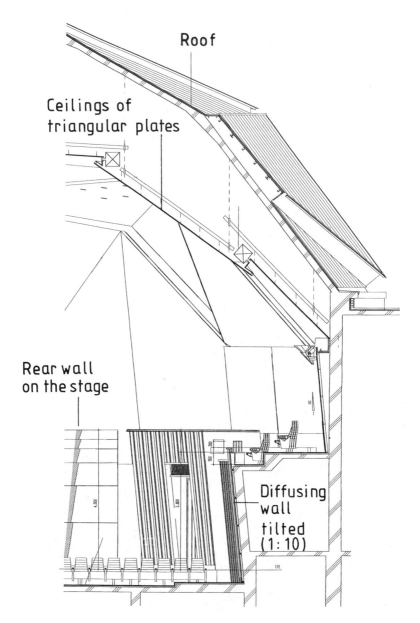

FIGURE 10.7. Details of the cross-section, including a sectional detail of the rear wall on the stage at the lower left.

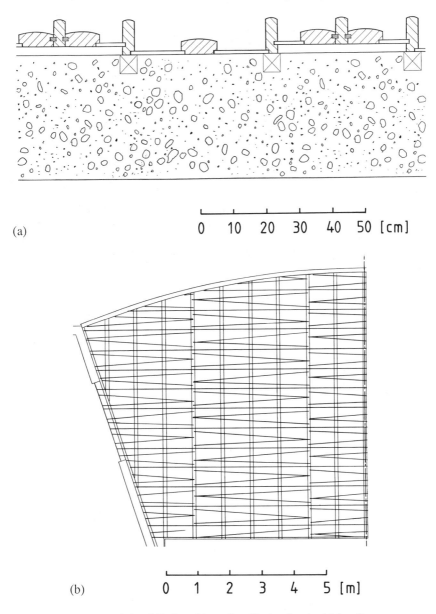

(a)

0   10   20   30   40   50 [cm]

(b)

0   1   2   3   4   5 [m]

FIGURE 10.8. Detail of the diffusing side walls effective for the higher-frequency range above 1.5 kHz, avoiding the image shift of the sound source well on the stage. The surface is deformed from the Schroeder's diffuser by removal of the partitions (a). Detail of the triangular joist arrangement for the stage floor, avoiding anomalous radiation due to the normal modes of vibration by some musical instruments touching the floor (b).

Photo 10.2. The Kirishima International Concert Hall, Kagoshima, Japan, with tilt side walls and triangular ceilings (photo by Sakimoto).

when the hall is designed for speech with $\tau_e = 22$ ms, then the reverberation time may be determined by maximizing the subjective preference centered on about 0.5 s ($\approx 23 \times 22$ ms, occupied). In the use of the auditorium for multiple purposes, however, the subsequent reverberation time must be controlled, conforming to the value of τ_e of the program source.

Acousticians have attempted to control the IACC, and the subsequent reverberation time by changing either the absorption coefficient of the boundary, the total cubic volume of the room, or both. In addition, some acoustical designers modified these factors may by an electroacoustic system, enhancing the sound field in the auditorium.

10.2.1. A Round-Shaped Hall

Since the time of ancient Greece and Rome, architects attempting to design circular-shaped auditoria have met with some acoustic difficulty. Nevertheless, for the purpose of some events and fashion shows, the round-shaped hall is designed like an "unidentified flying object (UFO)."

In order to control the acoustic factors, due to the subjective preference theory described in Chapter 4, a number of acoustic elements are considered, as shown in Figure 10.9(a, b) (Takatsu, Mori, and Ando, 1997; Takatsu, Hase, Sakurai, Sato, and Ando, 1998).

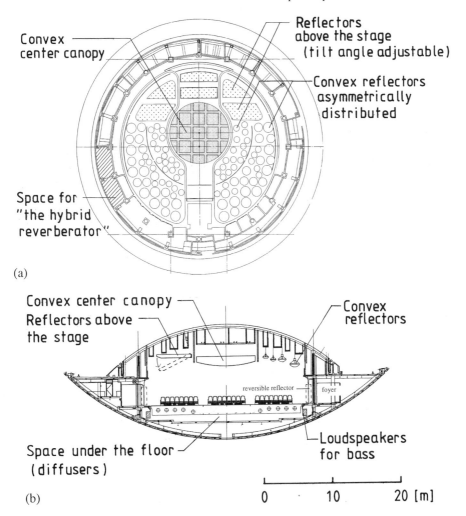

FIGURE 10.9. Detail of ceilings (a), and cross-section (b) of the round-shaped event hall for the Fashion Plaza, Kobe. A number of elements are designed for both of the left-hemispheric-temporal factors and the right-hemispheric-spatial factors.

(1) Reflectors above the stage are designed with adjustable tilt angles. The effects of the tilt angle on the values of the IACC are shown in Figure 10.10. The left-hand part of the figures indicates the contour lines of equal IACC values calculated with the horizontal reflectors, and the right-hand part indicates those with the appropriate tilt angle of the reflectors. Obviously, the tilt reflectors are effective in decreasing the IACC throughout the audience floor of 400 seats. Figure 10.11 shows the interaural cross-correlation function calculated for the 500 Hz octave band range. These results ensure $\tau_{IACC} = 0$, so that no image shift of the sound source occurs.

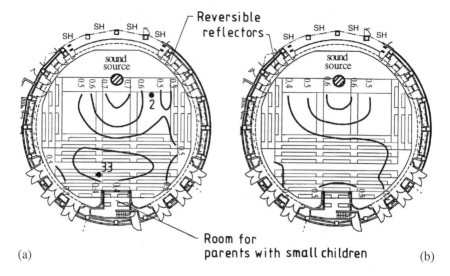

(a)   Room for parents with small children   (b)

FIGURE 10.10. Effects of reflectors above the stage on the IACC calculated at each seat. (a) Horizontal reflectors above the stage; and (b) reflectors tilted above the stage as shown in Figure 10.9(b).

(2) For the purpose of decreasing the IACC over a wide range of frequency, and based on the most suitable direction of reflections for each frequency band as shown in Figure 6.3, a number of convex reflectors are distributed asymmetrically, close to the ceilings. These are used as lighting covers as well. The smaller diameters of the convex reflectors are placed in the center of the ceiling to obtain reflections of a higher-frequency range (Photo 10.3), because the appropriate direction of reflection to listeners must be kept at a small angle measured from the median plane (above 2 kHz). The large diameters are effective for the lower-frequency range, providing large angles of reflection from the median plane (Section 8.3).

(3) The convex center canopy acts as a diffuser for avoiding strong reflection in the median plane, but providing enough energy to all the seats.

(4) The spaces both above and below the ears play equally important roles in the binaural and spatial factor (IACC). When the floor is acoustically transparent, then the space under the floor is used to avoid the strong low-frequency attenuation (Section 8.5), and at the same time is used to control the IACC with diffusers placed under the floor.

(5) The reversible reflectors with and without absorbing material on the side walls, on and near the stage, control the IACC as well as the T_{sub}.

(6) The room for parents with babies located at the side opposite the stage attenuates the long-path echoes along the circular wall in the hall.

(7) The small reverberant spaces with different reverberation times, together with the digital reverberation machines, form a hybrid reverberator. This system, with the loudspeakers distributed near the ceiling and under the floor, is designed

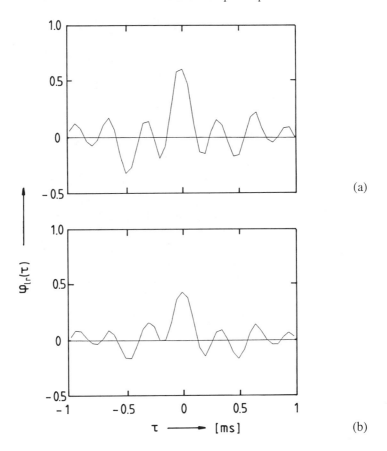

FIGURE 10.11. Examples of the interaural cross-correlation function calculated at seats 2 and 33 shown in Figure 10.10(a), with the data of Music Motif B.

to add reverberation time to the room was designed as 0.5 s, according to the type of program sources.

In a manner similar to the method of designing the concert hall with the music sources as well as the speech located on the stage, an electroacoustic system with a multiple channel of loudspeakers may also be designed. Taking the directivity of the loudspeakers, properties of the delay machines, and the reverberators and reflection properties of the walls into consideration, we can calculate the impulse response at each seat (Takeuchi, Mori, and Ando, 1997). Hence the global subjective preference, according to the four acoustic factors, may be obtained. Particularly, for public address systems, the range of effective duration of the ACF (τ_e) of speech signals may be 20 ms to 30 ms. And, for music reproduction, the range of τ_e should be decided depending on the purpose of the acoustic space. Usually, the range of τ_e for music is much greater than that for speech. Thus, two sound systems

PHOTO 10.3. Computer graphic for the round-shaped event hall for the Fashion Plaza, Kobe, Japan.

at least are recommended to be designed, for maximizing subjective preference. If the subjective preference of speech is maximized, speech clarity may also be increased by both of the temporal factors and spatial factors minimizing the IACC (Section 6.4).

10.2.2. A Hall with Movable Stage Towers

This example has been developed by ARTEC, New York, and illustrates how a large stage accomodating theatrical scenery may be transformed into a successful hall for musical performance.

(1) A number of movable stage towers weighing about 1 ton each may arrange the shape of the stage enclosure and its size according to the music program source. The area of the stage floor may also be changed by two elevators which are the floors of the orchestra pit as shown in Figure 10.12.
(2) The acoustical canopy above the stage is changed in height to provide back reflections for the performers (Figures 7.3 to 7.5) as well as the listeners (Figure 6.3).
(3) The tilt ceilings above the audience ensure small values of the IACC for a seat far from the stage.

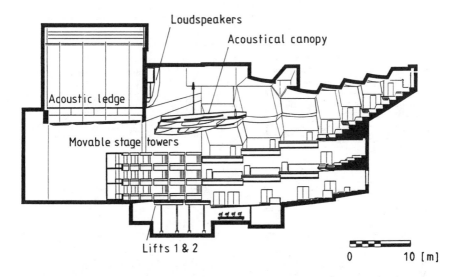

Figure 10.12. Cross-section of the El Pomar Great Hall, Pike's Peak Center, Colorado Springs, USA (after Architectural Record, August 1984). The acoustic canopy above the stage and the movable stage towers are arranged according to the music program.

10.2.3. A Hall with Variable Coupled Cubage

As shown in Figure 10.13, Johnson, Kahle, and Essert (1997) proposed a method to control the decay characteristics of the reverberation time in concert halls, incorporating partially coupled spaces and keeping a certain degree of clarity. This method includes both the variable couplings and the variability in design features, i.e., both the width and height of the hall and the strength of the early reflections. Thus, we can optimize the sound field to suit individual musical compositions. Figure 10.13(a) provides a narrow room with a low ceiling, and Figure 10.13(b) provides a room with a high ceiling with a width of 35 m. Examples of such reverberance chambers may be seen at the Festival Hall in Tampa, Florida, the Meyerson–McDermott Concert Hall in Dallas, the Crouse Hinds Concert Theatre in Syracuse, New York, and others.

These controls may be performed by a "sound coordinator," a specialist similar to a curator in the museum. The sound coordinator must understand the musical arts as well as acoustical science.

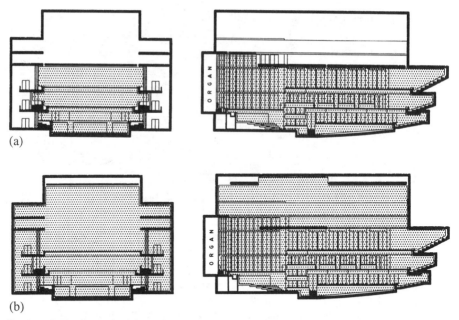

(a)

(b)

FIGURE 10.13. Control of the reverberation time and early reflections by change to the volume of the room (Johnson, Kahle, and Essert, 1997).

11

Acoustical Measurements of the Sound Fields in Rooms

Following construction, acoustic measurements are made for the purpose of test-ing the acoustic factors which were calculated at the design stage of the halls. The accumulation and understanding of such data improves the calculation of the acoustic factors performed at the design stage.

11.1. Binaural Impulse Response

A diagnostic system of measuring the impulse response at the two-ear entrances, determining the acoustic factors, and further evaluations of subjective attributes of the sound field at each seat in a hall, is shown in Figure 11.1. A pseudo-random-binary signal is radiated from the loudspeaker to measure the impulse responses by two tiny microphones placed at the two-ear entrances of a real head (1.1 m above the floor). Then, the spatial factors of the right hemisphere specialization (LL and IACC) and the temporal factors of the left hemisphere specialization (Δt_1 and T_{sub}) are analyzed. When the effective duration of the ACF of the source signal (τ_e) is analyzed, then the total scale value adding the scale values of the spatial factors $g_r(x)$, and of the temporal factors $g_l(x)$, referring to the most-preferred conditions, may be obtained. The value of $(\tau_e)_{\text{min}}$ is used to determine the most-preferred temporal values for $[\Delta t_1]_p$ and $[T_{\text{sub}}]_p$ (Sections 3.1 and 4.3). The scale value of the subjective preference of the sound field is obtained after obtaining the measured physical factors. If the source signal is fed into an ACF-processor, then outputs numbered 1 through 4 may be used to control the sound field with an electroacoustic system simultaneously without any manual adjustment, preserving the preferred conditions of the four factors.

The pseudo-random-binary signal m_j is generated by a system with shift-registers as shown in Figure 11.2 (Davies, 1966). For example, if $n = 4, k = 3$, and the initial shift-registers are -1, then the output sequence becomes

$$m_j = -1, -1, -1, -1, +1, +1, +1, -1, +1, +1, -1, -1, +1, -1, +1$$
$$(j = 1-15).$$

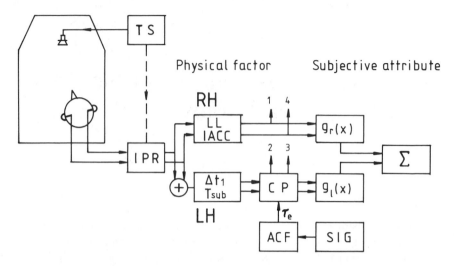

FIGURE 11.1. A system of measuring the four orthogonal factors and evaluating subjective qualities at each seat in a room. TS: test signal (maximum-length-sequence signal); IPR: impulse response analyzer; RH: right hemispheric factors (listening level and IACC); LH: left hemispheric factors (Δt_1, T_{sub}, and the A-value in addition); CP: comparators with the most-preferred conditions based on the effective duration of ACF, τ_e; ACF: autocorrelation-function processor, τ_e; SIG: source signals; $g_r(x)$: scale values from the right hemispheric factors; $g_l(x)$: scale values from the left hemispheric factors; and S: total scale value of a certain subjective attribute.

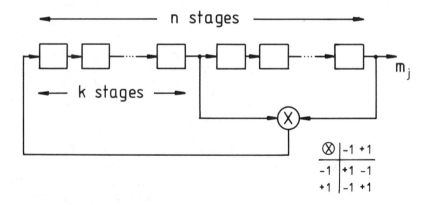

FIGURE 11.2. Generation of a maximum-length sequence (Davies, 1966).

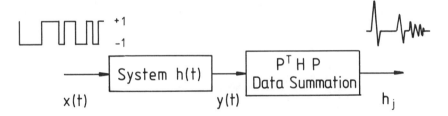

FIGURE 11.3. A fast method of measuring the impulse response (Alrutz, 1981).

This is repeated with a period of 15 binary digits. The largest possible period for the system is given by $L = 2^n - 1$ (in this case, $n = 4$), and is called the maximum-length sequence.

The input signal $x(t)$ to the linear system $h(t)$ under test is obtained by the sequence m_j. As shown in Figure 11.3, the algorithm $P^T H P$ enables us to compute the impulse response with only a summation of the output data $y(t)$ from the system (Alrutz, 1981). Since the Silvester-type Hadamard matrix H contains either 0 or 1, the computation is performed by adding operation without any multiplication operation.

Examples of measuring the binaural impulse responses at a seat close to the stage (Seat a, left ear) and at a rear seat b (right ear) in the Kirishima International Concert Hall are demonstrated in Figure 11.4. In this measurement, an omni-directional-dodecahedron loudspeaker with 12 full range drivers was placed on the stage 1.5 m above the floor for a sound source. It can be observed that the total amplitude of reflections, A, at seat a (close to the stage) is smaller than that at seat b (far from the stage).

11.2. Reverberation Time

After the impulse response is obtained, the reverberation time is measured by the Schroeder method (1965a, b). The integrated decay curve as a function of time may be obtained by squaring and integrating the impulse response of the sound field in a room, such that

$$\langle s^2(t) \rangle = K \int_{t+T}^{t} h^2(x)\, dx, \tag{11.1}$$

where the time T should be chosen sufficiently longer than the reverberation time.

Examples of the decay curve measurement and the decay rate of both the left and right ears at seat a for the 500 Hz-octave band are shown in Figure 11.5. The reverberation times measured are both 2.07 s. The measured reverberation times with octave band filters in the Kirishima International Concert Hall (without an audience) are plotted as full circles as shown in Figure 11.6. The empty circles are estimated values of the reverberation time for a full audience.

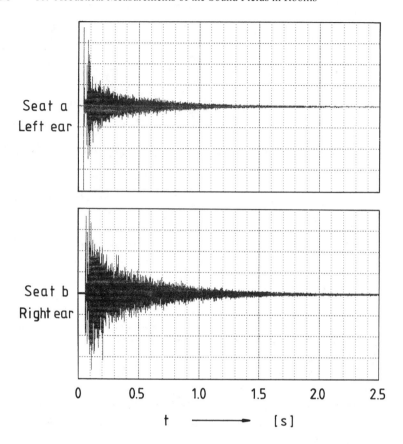

FIGURE 11.4. Impulse responses measured at seats a (near to the stage) and b (far from the stage) in the Kirishima International Concert Hall. The amplitude of the impulse response measured at seat a is attenuated.

It is worth noticing that Jordan (1969) showed that the values of the early decay time (EDT) measured over the first 10 dB of decay are close to the values of the reverberation time averaged with the interval of −5 dB to −35 dB.

The total amplitude of reflection A, defined by Equation (4.6), is obtained as its square

$$A^2 = \frac{\int_{\infty}^{\varepsilon} h^2(x)\,dx}{\int_{\varepsilon}^{0} h^2(x)\,dx},\tag{11.2}$$

where ε signifies a small delay time just large enough to cover the duration of the direct sound.

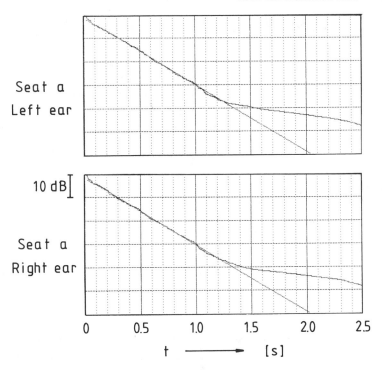

FIGURE 11.5. Integrated decay curves obtained from the impulse responses at the two ear entrances, at seat a in the Kirishima International Concert Hall.

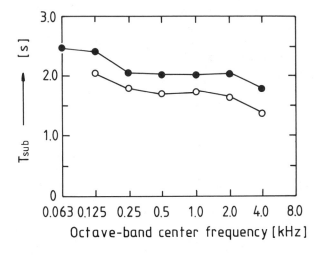

FIGURE 11.6. Reverberation time measured in the Kirishima International Concert Hall. (●): measured values without audience; and (○): estimated with full audience.

11.3. Measurement of Acoustic Factors at Each Seat in a Concert Hall

We discussed in Chapter 9 the seat selection system designed for the purpose of enhancing individual satisfaction. To begin with, four orthogonal factors are measured at each seat in a concert hall (Ando, Sato, Nakajima, and Sakurai, 1997; Nakajima and Ando, 1997; see also Sakurai, Aizawa, and Ando, 1998).

Measured values of the listening level (LL), the total amplitude of reflection (A), the initial time delay gap (Δt_1) between the direct sound and the first reflection, excluding the reflection from the floor, and the IACC at each seat are shown in Figure 11.7. The reverberation times at all the seats had almost the same value, about 2.05 s for the 500 Hz band.

Even though the final scheme of the concert hall was changed in width of the hall (1 m larger) from the scheme at the design stage, the values of each physical factor measured, as shown in Figure 11.7, are not much different from the values calculated in Figure 10.4.

11.4. Recommended Method for the IACC Measurement

The following two methods are recommended for measuring the IACC, as needed for subjective evaluations or for the sound field tests after construction of a room:

(1) In order to evaluate the subjective responses of the sound field, the interaural crosscorrelation measurement (with all values of the IACC, τ_{IACC} and W_{IACC} defined in Figure 3.7) is recommended to be performed. These measurements are performed after passage through the A-weighting network with the music or speech signal, under identical conditions with subjective judgments at each seat position in an existing hall or at the listening seat of a simulated sound field in the laboratory. If a loudspeaker is used as a source signal in the room under test, then the same characteristic source signal must be used both in the measurement of the physical acoustic factors and in the subjective tests. Since the interaural cross-correlation function is a spatial factor, it is recommended that measurements be made for all of the direct sound and reflections without any temporal subdivisions. It is worth emphasizing that the preferred condition is $\tau_{IACC} = 0$, as described in Section 4.5.

(2) In order to compare the sound field in the existing hall after construction with that calculated at the design stage, measurements of each octave-band are performed, together with the other factors, Δt_1, T_{sub}, A and LL. This may include the measurements of the $IACC_E$ (Hidaka, Beranek, and Okano, 1995), which is defined by 0 ms to 80 ms of the integration interval for the direct sound and early

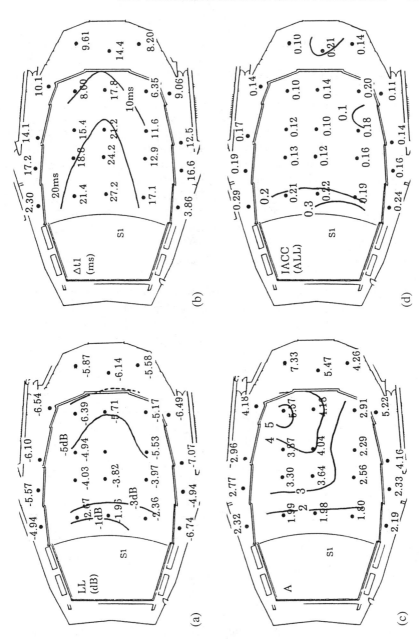

FIGURE 11.7. Physical factors measured at each seat in the Kirishima International Concert Hall. (a) Listening level; (b) Δt_1; (c) A value, the total amplitude of reflections; and (d) IACC.

reflections, as well as the IACC, which is defined by 0 ms to $-\infty$ ms for the direct sound and all reflections and reverberation. A typical example of measuring the IACC as a function of the integration interval, which was performed in Symphony Hall, Boston, is shown in Figure 11.8. It is remarkable that a close relationship is found between the values of the $IACC_E$ at 80 ms and the IACC (at 3000 ms) as shown in Figure 11.9.

When the space is used for performing a dance, ice skating, or a party, then the listeners are facing various directions. Then values of the IACC and τ_{IACC} are measured as a function of the direction of the head. The measured results with the 500 Hz octave-band noise in an oblong atrium of a hotel at a distance 10 m from the source position are demonstrated in Figure 11.10. When the listener is facing the sound source, then IACC = 0.41, and τ_{IACC} = 0, and thus no image shift occurs. These values are nearly unchanged for the head directional angles less than 30°. When the listener is facing the lateral side at 90°, then the IACC became greater than 0.50, and τ_{IACC} is about 600 μs, due to the "lateral sound source."

If we are interested in the "apparent source width (ASW)," then the calculation and measurement of W_{IACC} of the sound field are recommended. (The calculation may be carried out by Equation (3.25), with the value of δ defined as shown in Figure 3.7.)

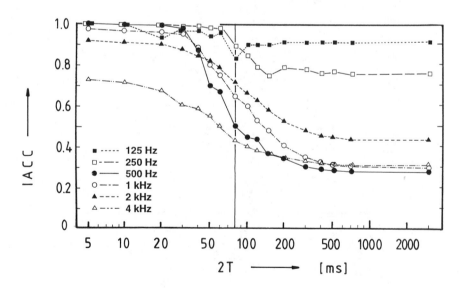

FIGURE 11.8. Measured IACC as a function of the integration interval, $2T$, of the impulse responses, for each octave-band range (Hidaka, Beranek, and Okano, 1995).

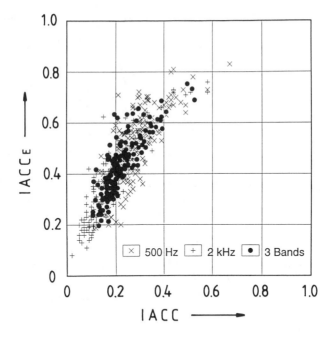

FIGURE 11.9. The relationship between the values of the IACC$_E$ and IACC (Hidaka, Personal communication, 1996).

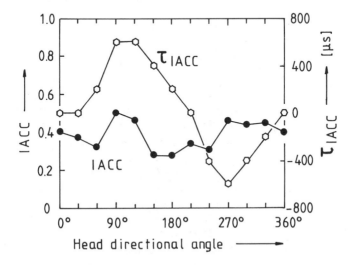

FIGURE 11.10. Measured IACC and τ_{IACC} as a function of the head direction to the sound source, in a hotel atrium (Kobayashi et al., 1997).

11.5. Physical Properties of a Forest as an Acoustic Space

It is believed that birds singing in the forest was the origin of music, and the forest was probably the first performance space of vocal music during a walk. The sound field in the forest consists of multiple scattering phenomena, and it is complicated to calculate the impulse response even by the use of a computer. Thus, the impulse-response measurement was made by reproducing the pseudo-random-binary signal of 2.7 s with the clock frequency of 48 kHz. The knowledge obtained here may be utilized to understand music performance in the forest, and further to design the sound field in rooms with a large number of columns.

In order to investigate the effects of multiple scattering by trees, acoustic factors were measured on a 5 m wide path of asphalt in a forest as is shown in Photo 11.1 and Figure 11.11 (Sakai, Sato, and Ando, 1996; 1998). The sketch shows that, within a part of the forest with an area of 60 m × 45 m, there were trees of various diameters between about 0.3 m and 1.0 m; the average height of the trees was about 18.5 m. The omnidirectional-dodecahedron loudspeaker with 12 full-range drivers was placed 1.5 m above the path, and the pseudo-random-binary signal was radiated from a loudspeaker similar to that in the previous section. The impulse responses obtained at the two-ear entrances with a real head (20 m from the source position, 1.6 m in height) are shown in Figure 11.12. The strong reflection from the asphalt surface at about 1 ms is observed, but no clear initial time-delay gap between the direct sound and the first reflection from the trees is identified. Thus, three of the four factors were measured here. The integrated decay curve obtained from the impulse response at the left ear entrance is demonstrated in Figure 11.12(c). The logarithmic decay may be found for the initial decay range of about 15 dB, just after the strong direct sound and the asphalt reflection. The typical multiple scattering effects may be observed in the curve after 0.4 s departing from the logarithmic decay. This decay curve greatly differs from the curves shown in Figure 11.5 which were measured in the room.

The sound-pressure levels for each octave-band signal as a function of the distance relative to the pressure level at 5 m are shown in Figure 11.13. It is of interest to note that the level of the 125 Hz band signal was higher than the inverse square law at 10 m to 20 m, due to the reflection of the asphalt surface, while the sound-pressure level for other frequency bands decreased more than the law, due to the long path multiple scattered reflections and absorption by the trees.

The reverberation times as a parameter of distance are shown in Figure 11.14. The remarkable findings here are that:

(1) Up to 5 m from the source, the reverberation time is shorter than 0.7 s. The longest reverberation time is observed at 500 Hz.
(2) The reverberation time increased with the increasing distance from the source, particularly for values in the middle frequency range of 500 Hz where it rose to about 1.8 s at a distance of 40 m.

Photo 11.1. A forest investigated for its acoustic factors.

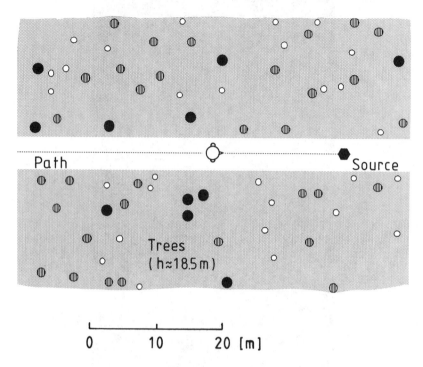

FIGURE 11.11. A forest where the orthogonal acoustic factors were measured. The diameters of trees distributed are roughly classified by 0.3 m, 0.6 m, and 1.0 m.

(3) The reverberation time in the lower-frequency range (below 250 Hz) is shorter than 1.3 s. However, the listening level is relatively higher than the higher-frequency range creating loudness balance.

The measured IACC are shown in Figure 11.15. For the frequency range above 1 kHz, the value of the IACC rapidly decreased with distance. The IACC measured for such a frequency range at a distance of 40 m was less than 0.5.

Referring to these measured factors (except for the initial time-delay gap), it is concluded that the sound field in the forest is an excellent acoustic condition for sources such as flutes and string music, and vocal music as well as for bird singing which contain the middle frequency components between 500 Hz and 2 kHz. This is most true, particularly at a distance about 40 m from the source location.

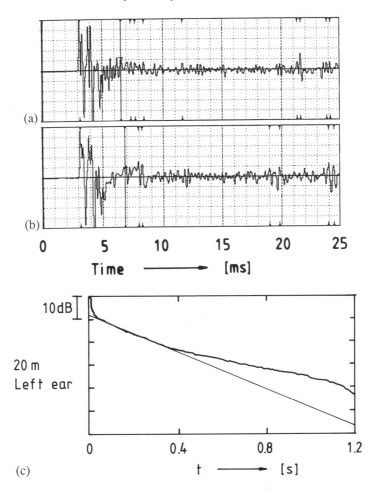

FIGURE 11.12. Binaural impulse responses measured at 20 m from the sound source. (a) Left ear entrance; (b) right ear entrance; and (c) integrated decay curve at the left ear entrance.

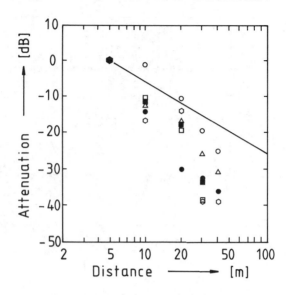

FIGURE 11.13. Relative LL measured as a function of the distance. (○): 125 Hz; (△): 250 Hz; (●): 500 Hz; (□): 1 kHz; (■): 2 kHz; and (◯): 4 kHz.

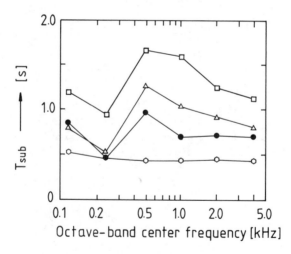

FIGURE 11.14. Measured reverberation time as a parameter of the distance from the sound source. (◯): 5 m; (●): 10 m; (△): 20 m; and (□): 40 m.

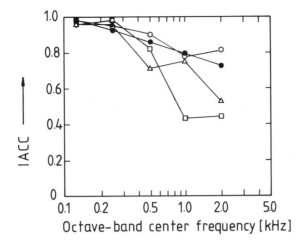

FIGURE 11.15. Measured IACC as a parameter of the distance from the sound source ($\tau_{IACC} = 0$). (○): 5 m; (●): 10 m; (△): 20 m; and (□): 40 m.

12

Generalization to Physical Environmental Planning Theory

We have evolved in a universe which can be measured by using the dimensions of time and space. Consequently, the dimensions of the "inner universe"—an individual's subjective model of the universe—must correspond in some way to those of the physical universe. The brain receives environmental stimuli not only from the hearing system, but also from visual, thermal, and other senses, to some extent or another. These are associated with an individual's total subjective and physiological evaluation of any physical space and more particularly in a concert hall.

This chapter suggests a foundation for the theory of planning physical environments, which takes account of time and space as they are specialized in human cerebral hemispheres. A representative example of the application of this concept of environmental planning has been discussed in the chapters on concert hall acoustics: the sound field in a hall can be altered with careful manipulation of four orthogonal factors. These variables comprise two temporal factors of the left-hemisphere dominance, and two factors of the right hemisphere dominance which are related to spatial attributes. The design of a specific concert hall can be altered to suit specific types of music, such as chamber music or choral works. This concept can be partially found in the architecture of Tadao Ando (Frampton, 1985; 1987), the traditional tea houses in Japan, and the urban design theory of Lynch (1972).

The physical variables are specific, measurable factors, such as, for example, the temperature and lighting level, that influence a person's perception of the environment. If these variables are identified, and the interrelationships and influences on human perception explored, a method can be evoked that has application in the field of physical environmental design, including architecture and urban design.

12.1. A Generalized Theory of Designing Physical Environments

12.1.1. Hemispheric Specialization

It has been noted that the two hemispheres of the brain have different areas of specialization and ability. As discussed in Chapter 5, subjective preference about time-factored experience for the sound field takes place in the left hemisphere, and the spatial-factored experience in the right hemisphere (Ando, Kang, and Morita, 1987; Ando, 1992; Ando and Chen 1996; Chen and Ando, 1996; Nishio and Ando, 1996; Chen, Ryugo, and Ando, 1997). The right hemisphere tends to perceive space in multiple-dimensional, nontemporal, and nonverbal terms for both visual and auditory environments. In most subjects, the left cerebral hemisphere is normally concerned with linear, sequential modes of thinking, such as speech and calculation (Sperry, 1974; Davis and Wada, 1974; Galin and Ellis, 1975; Levy and Trevarthen, 1976; Ando and Kang, 1987).

Clearly, most experiences in daily life involve a combination of these two modes of thought, with instantaneous dominance of the left and right switching continually; and both hemispheres often cogitating simultaneously. It is similar to photographs or portraits of being a two-dimensional representation of reality. When the time function is added to a projection of a sequence of these images, the static image "springs to life."

Imaging increasing levels of complexity, we have built up a list of factors (Table 12.1). These account for the physical orthogonal factors and influence everyday human experience. An ideal value for each factor, or a range of preferred values, can

TABLE 12.1. Proposed physical environmental factors to be planned.

Physical environment	Spatial factor	Temporal factor
Acoustic	(1) Listening level, LL	(2) Initial time delay gap between the direct sound and the first reflection, Δt_1.
	(4) Interaural cross correlation, IACC	(3) Subsequent reverberation time, T_{sub}
Visual	(1) Lighting level	(2) Properties of movement function of reflective surface, T
	(3) Properties of the reflecting surface	
	(4) Spatial perception including distance factor	
Thermal	(1) Spatial sensation of body	(2) Relative humidity
		(3) Temperature
		(4) Radiant Heat
		(5) Air movement (velocity)

Note: In addition to the above variables, the characteristics of sources, the location change of source, and the observer's activity should also be taken into consideration.

be found by systematic research utilizing the theory of subjective preference with a number of subjects. A matrix can thus be assembled that describes the ideal values and the mutual influence or independence of each. Such a planning tool could be utilized during the design phase of a project to predict whether humans will find a given physical environment pleasurable, before it is built. What such a planning tool would add to the practice of environmental design is a method that explicitly recognizes the needs of the specialization of the human brain. Specifically, a design method is proposed that appreciates the sensitivity of the left hemisphere to the dimension of time in the environment, in addition to the common consideration of three dimensions of space.

There have been notable successes in the attempt to introduce consideration of a time element, specifically in terms of the sequential nature of movement through space, into the process of design. What distinguishes the present approach from previous ones is its attempt to describe an environment in terms that include the passage of time together with a spatial factor. Previous studies have focused primarily on an individual's movement through space (Figure 12.1).

12.1.2. Designing Physical Environments

A general theory for evaluating subjective attributes is described, based on the theory of subjective preference for a sound field (Section 4.4).

Let x_n, $n = 1, 2, \ldots, I$, be the significant physical factors of I dimensions acting on a human; then the scale value of subjective preference, or another single

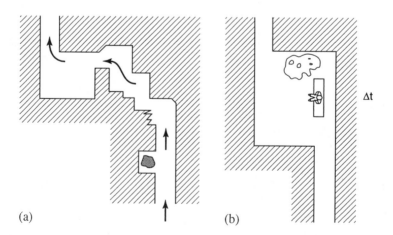

FIGURE 12.1. Sequential versus temporal experience of the environment. (a) Movement through space previously considered; and (b) experience of the passage of time while stationary.

subjective response, is given by

$$S = g(x)$$
$$= g(x_1, x_2, ..., x_I). \tag{12.1}$$

Next, we consider the fact that the function of the human cerebral hemispheres can be divided into two main categories: spatial factors which are associated with the right hemisphere, and temporal factors which are associated with the left. Thus, we can divide the physical factors defining human perception of the environment into two, so that Equation (12.1) may be reduced to (Ando, Johnson, and Bosworth, 1996)

$$S = g_l(x) + g_r(x)$$
$$= g_l(x_{l1}, x_{l2}, \ldots, x_{lM}) + g_r(x_{r1}, x_{r2}, \ldots, x_{rN}), \qquad I = M + N. \tag{12.2}$$

Furthermore, considering the fact (as mentioned in Section 9.2) that there are few effects of lighting level on the preferred initial time-delay gap between the direct sound and the first reflection, minor interference is likely observed between different physical factors such as between the visual factor and the auditory factor. For instance, as described in Sections 5.2 and 5.3, there is little interference since the initial time-delay gap is associated with the left hemisphere and the lighting level is assumed to be the right-hemisphere dominance.

It is assumed that each physical environment has an independent influence on the subjective attributes from other physical environments, so that

$$S = [g_l(x) + g_r(x)]_{\text{auditory}} + [g_l(x) + g_r(x)]_{\text{visual}} + [g_l(x) + g_r(x)]_{\text{thermal}}$$
$$+ [g_l(x) + g_r(x)]_{\text{other human physical environments}}. \tag{12.3}$$

In particular, this holds, at least in the neighborhood of the optimal conditions for each physical factor, avoiding an extreme physical environment, for example, listening to music in a temperature below $0°$ C.

If physical environments other than the sound field are fixed at or near the optimal conditions, then

$$S = [g_l(x) + g_r(x)]_{\text{auditory}} + c, \qquad I = 4. \tag{12.4}$$

This is the same formula as discussed in Section 4.4. Without loss of information in this expression, due to the fact that the scale is a relative one, we can put the constant at $c = 0$.

Since there is a lack of data calculating all of the effects in terms of the scale value, the last three terms in Equation (12.3) are, as yet, unavailable. However, we can be aware of the significant factors in designing physical environments. In order to blend human and physical environments, both the spatial and temporal factors must be taken into account, similar to the theory described in concert hall acoustics, blending listeners and the sound field together with the musical tempo.

In architectural planning, we are likely to forget the temporal factors which must be included in the plan and cross-sections of the building. Figure 12.2 indicates discrete temporal periods of the human and physical environments which have to

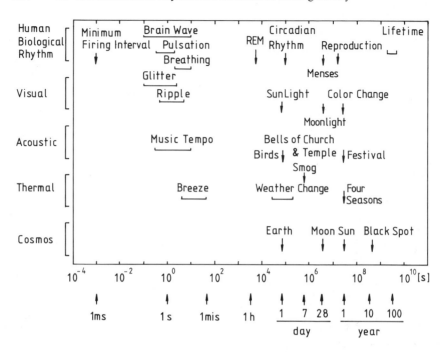

FIGURE 12.2. Blending human biological rhythms and periods of natural physical environments.

be blended. Remarkably, there are certain significant periodic eigenvalues, in both human biological rhythm and physical environmental activities in the time domain, to be blended. Therefore, we do not need to consider every continuous-time period with the possible infinite-real number.

It is worth noticing that there are many more unconscious physiological rhythms than subjective psychological attributes, associated with the physical environmental activities of long periods. For example, the effects of environmental noise are described in terms of the development of unborn babies and the development of children over long period accumulations: periods of reproduction and generation (Ando and Hattori, 1973, Ando, Nakane, and Egawa, 1975, Ando and Hattori, 1977a, b; Ando, 1988). The following section discusses the significant factors for each physical environment for the conscious perception of the psychological present.

12.1.3. Proposed Factors for Physical Environments

In the search for a broader approach to design, the same approach used to analyze acoustic performance has been applied to other realms of human experience, namely, the visual and thermal.

(A) Visual

In a manner similar to acoustics, human perception of the visual environment can be expressed by using Equation (3.14), which describes a visual image at certain points of the left and right retina

$$f_{l,r}(t; R_0) = \{p_n(t; R_0) * A_n w_n(t, T) * h_{nl,r}(t)\}, \tag{12.5}$$

in which $p_n(t; R_0)$ signifies the characteristics of a light source located at $R_0 = (x_0, y_0, z_0)$; n is an integer indicating each ray of light; A_n is the distance attenuation according to the inverse law; $w_n(t)$ signifies the properties of the surface from which light is being reflected; T is the time representing the movement of the surface; and $h_n(t)$ is the physical characteristic of the "spatial perception system" including the eyes and face of an observer located at $R = (x, y, z)$, $\{\}$ designating a set.

If we rewrite the left side of Equation (12.5) to take account of the movement of the observer, it becomes

$$f_{l,r}\left(t; R(t)|R_0\right), \tag{12.6}$$

where $R(t)$ signifies the observer's movement as a function of time t. This equation, therefore, has both spatial and temporal constituents. There is the spatial relationship of the light source and the observer; and a time dimension in the movement of the observer through space. Thus the visual environment can be described using the following factors, other than light source characteristics, and the movement factor $R(t)$ of the observation:

(1) lighting level;
(2) the time factor T of a reflector;
(3) properties of the reflecting surface, $w_n(t)$, which include color and form (edge); and
(4) properties of the spatial perception, $h_{nl,r}(t)$ and the distance factor A_n;

(B) Thermal

Sensation of the thermal environment can also be described using the following factors:

(1) spatial distribution sensed by the whole body;
(2) relative humidity;
(3) temperature;
(4) radiant heat; and
(5) air movement (velocity).

Common sense indicates, however, that thermal comfort is not a static function. For example, after being outdoors on a cold day, a person entering a department store experiences a temporary sensation that the store is "too hot," when, in fact, the temperature is at a normally comfortable level (Kuno, Ohno, and Nakahara, 1987). The thermal conditions that a person finds comfortable for riding an exercycle may be too chilly for sitting quietly and reading. Thermal comfort is thus conditioned by activity and the boundary conditions encountered. Factors for time and movement must be introduced.

Considered in this way, each environment's thermal suitability can be adjusted to the activities taking place therein, in a manner similar to that discussed for architectural acoustics. A kitchen, for example, should have a cooler thermal environment than a library, because the kitchen has a higher level of human activity. Table 12.1 indicates suggested environmental variables. These factors are limited to those that have the possibility of being altered by designers.

12.2. Examples of Physical Environmental Planning

12.2.1. Discrete Periods of Environment and Human Life

As shown in Figure 12.2, the crucial factor in the temporal dimension of the environment is the periodic cycle. Every aspect of the passage of time is bound up with periods: for example, the shortest period (about 0.5 s to 5 s) corresponding to the psychological present is related to brain wave, pulsation, and breathing, which are associated with the perception of music, the glitter of leaves, and the ripple of a water surface. The rapid eye movement (REM) of about 70 to 150 per minute, related to a basic rest-activity cycle of 10 to 20 per day (Othmer, Hayden, and Segelbaum, 1969; Kripke, 1972), is associated with, for example, one session of a concert, a lecture, and work. The circadian rhythm deeply connected by sunlight with the Earth's rotation period is associated with daily human activity. The week created by the social law for work and leisure is associated with, for example, the planning period of concert and drama, or a social activity. The next distinguishable period is concerned with the movement of the Moon. The revolution of the Earth around the Sun (changing of the seasons) is associated with, for example, the color change of leaves and annual festivals. The black spots on the Sun which appear once in about every 11 years may influence, more or less, the environments on Earth. Such a periodic space weather, including the magnetic storm may cause effects on human life (Roederer, 1995). The alternation of generations of about 30 years and the span of life of, say, about 90 years, may be considered in the planning of houses, in accordance with the individual schedule of life.

The present theory suggests that these discrete periods should be explicitly recognized during the design process for any human environment. The passage of time in the designed environment should be as consciously considered as the three-dimensional organization of the space itself. The following are examples of designing physical environments related to a concert hall.

12.2.2. Physical Environments for a Concert Hall

(A) Approach

The approach passage to a concert hall is designed for both temporal and spatial factors, so that our senses will gradually be excited in anticipation of the music.

In the design of lane, for example, trees, soil, water, fire, Sun, Moon, and stars help to produce music in our mind escaping from our daily life to the concert. Direct ways of producing sound are birds singing in trees, leaves in a breeze, and a waterfall. Indirect ways of producing music in our mind are due to quiet elements, stars, Moon, Sun, a skyline of mountains. Thus, just before the concert begins, a well-designed sound signal should be reproduced to inform audience to be seated.

(B) Visual Environment

In order to design a good visual environment, both for the musicians and listeners in a concert hall, there are several factors to be controlled, as indicated in Table 12.1. Relating to the lighting level, Claus Ocker (1995) as a singer suggests that nonverbal communication between the musician and the listener in a concert hall, as shown in Figure 12.3, is full of emotion and tension on both sides. Communication starts at the moment when the door to the stage is opened. The performer would like to see the faces of the audience and to make contact with them during the performance, in order to recognize the audience's mood and to feel its intensity. Thus, more light in the hall than is customary is recommended for seeing the faces of the listeners seated in front of the musician. Ocker wonders about planed seats behind the stage, such as a "vineyard" as in the Berliner Philhamonie, because singers have the feeling that the sound of their voices is given in front of them, and not backward. An important part of artistic interpretation is that, in singing in an expressive way, a facial mimic can sometimes be seen in accordance with the sound of the voice coming from the deepest universe of soul or heart. This is an example of a special kind of communication between artists and audience.

Evidence that interaction between auditory information and visual information in performing percussion increased communication between the performers and listeners, as reported by Ohgushi and Sakuma (1997). For the visual environment, therefore, it is reconfirmed that the lighting level be about 35 lx is desirable, as was discussed in Section 9.2.

(C) Thermal Environment

The thermal environment is also important. Music performances, with low-level body activity and lighting on stage, need to maintain certain values of temperature and humidity. But many musicians are sensitive to unwanted fresh air from above

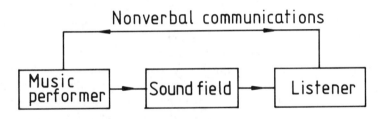

FIGURE 12.3. Nonverbal communications between music performer and listener.

onto the stage, moving musical note pages that have been turned or are in the process of being turned.

In being seated for a long time the knees, legs, and feet of the listeners are subject to become numb. This is a typical example of the spatial sensation controlling the distribution of temperature and air movement.

(D) Noise

It is well known that any possible noise must be kept to very low levels, hopefully, less than NC-15 (Beranek, 1957). Even the weak sound-pressure level of the electronic "peep-peep" sound of wristwatches or from portable telephones may suddenly interrupt the full concentration of the singer (Ocker, 1995). Continuous low-level sound from ventilation systems is likely to be ignored without any disruption, because of the right-hemispheric dominance, but the intermittent tonal peep-peep sounds influence our brain more critically than continuous noise because of additional information interference effects with music in the left hemisphere.

12.2.3. Play Area for Children and Residential Design

(A) Play Areas

To take another simple example of the application of this theory, it is easy to see why the play-space designs of recent years hold so much more interest than the "playsets." Wood structures, flexible bridges, and fanciful designs allow for much freer play of the imagination than the hard surfaces and steel frames of playground toys. It is also easy to see why natural areas such as small streams are so fascinating to children. The stream is an interactive "toy," changing course and speed as the child performs small-scale engineering projects, blending the children and water in the flow of time. The "toy" itself changes, due to the child's actions, and indeed continues to change even after the child stands back to admire his or her work. Other environmental elements include trees, soil, and flame, as well as celestial bodies such as the Sun, the stars, and Moon.

(B) Residential Design

Architectural form can be modified so that the natural periods of day and night, as well as clouds, sun, rain, and the seasons are embraced and made part of the experience of the enclosure (Ando, Johnson, and Bosworth, 1996). Architecture can blend and fuse natural cycles with human lives so that it becomes impossible to separate the two (Bosworth, 1997). Memory of a well-designed building bears with it, inextricably, a memory of spring sunlight as it shines across a white wall, or the sound of rain falling on the pavement outside.

There are three specific architectural devices that help bring these natural cycles of light and dark into a building. The first, looking like a proscenium of the auditorium for Nature's performance, is the window. The most obvious window type is a view window, with its top above and its sill below eye level. This allows people inside the building to see directly the changes in Nature taking place outside. The second type of window is a transom, located just above the human zone. This type brings light into the building and allows the inhabitants to see sky changes and other environmental phenomena, such as trees waving in the wind. The third window type is the monitor. Monitor windows are located above the habitable zone. Their use is more for bringing light in than for seeing it out. They allow a ceiling to be bathed in sunlight, or a rectangle of light to move across a wall during the course of the day.

The second architectural device is the courtyard (as an acoustic space, it is described in Section 7.3). It provides a closed view from within the interior space. This allows people to see natural periods of light and weather in a bounded portion of Nature with the architecture beyond.

The third type of architectural device is a garden that can be viewed from inside a building. It allows a controlled view of a portion of Nature with natural features visible beyond.

Architecture can intensify the sense of the passage of time by carefully using the devices. For example, light from a small window is magnified if the window is perpendicular to an adjacent wall. As the Sun moves, a ray of light plays across the wall like a sundial. As clouds pass, the room goes from light to dark and back again.

Interior surfaces can be modeled in the following ways:

(1) windows can be placed next to the ceiling, wall, and floor surfaces to reflect light; and
(2) the surfaces adjacent to the windows, transoms, or monitors can be splayed.

The effect is to magnify the light coming through the window. The same effect can be achieved with monitors next to a sloping ceiling. It is worth noting that this kind of system with a sloping ceiling creates a superior sound field, decreasing the IACC as is mentioned in Sections 8.1 and 8.2.

Another very pleasant lighting effect with seasonal variations can be achieved by orienting windows toward deciduous trees. A gable window, for example, will allow the trees to filter and reflect the incoming light. A similar effect comes from sunlight playing on water-reflecting onto a ceiling from the ocean, for example. Even while listening to music, these examples can be thought of as an invitation to Nature to enter and participate in the formation and experience of interior space.

For the purposes of blending physical environments and the human brain, the basic design theory has been developed incorporating temporal and spatial values. This concept was first developed scientifically for concert hall design as described in the previous chapters. The temporal and spatial factors strongly influence any subjective attributes and must be considered in the design of any physical environment. In the temporal design of an environment, for example, temporal rhythms in physical environments have to be blended with human biological rhythms.

So far, we have discussed that a meeting place of art and science may help discover individual preference or personality as the minimum unit of society. A lasting peace on earth may be achieved by release of each personality given by Nature.

Appendix I

Method of Factor Analysis

The method which is applied in the multiple-dimensional-factor analysis of Section 4.5 is briefly described here (Hayashi, 1952; Hayashi 1954a, b). We give the numerical values to each subcategory of each item and synthesize the responses as we are concerned with behavior patterns.

In this analysis, all items do not need to be scalable. Use the data of n cases. Let A be an outside variable and define s and k as $s = 1, 2, \ldots, R$ (R is the number of items), and $k = 1, 2, \ldots, K_s$ (K_s is the number of subcategories in the sth item), respectively. Since each case checks only one subcategory in each item, the behavior pattern of the i-case is to be synthesized in the form of

$$\alpha_i = \sum_{s=1}^{R} X_{s(i)} = \sum_{s=1}^{R} \left\{ \sum_{k=1}^{K_s} \delta_i(sk) X_{sk} \right\}, \tag{A.1}$$

where

$$\sum_{k=1}^{K_s} \delta_i(sk) = 1$$

and

$\delta_i(sk) = 1$ if the i-case comes under the kth subcategory in the sth item,

$\delta_i(sk) = 0$ otherwise.

α_i, which is called the total score of the i-case, has a numerical value, since X_{sk} has a numerical value.

The correlation coefficient ρ between A and α_i is written as follows:

$$\rho(A, \alpha_i) = \frac{(1/n) \sum_{i=1}^{n} (A_i - \bar{A})(\alpha_i - \bar{\alpha})}{\sigma_A \sigma_\alpha}, \tag{A.2}$$

where

$$\bar{A} = \frac{1}{n} \sum_{i=1}^{n} A_i, \qquad \sigma_A^2 = \frac{1}{n} \sum_{i=1}^{n} (A_i - \bar{A})^2,$$

$$\bar{\alpha} = \frac{1}{n} \sum_{i=1}^{n} \alpha_i, \qquad \sigma_\alpha^2 = \frac{1}{n} \sum_{i=1}^{n} (\alpha_i - \bar{\alpha})^2.$$

In order to obtain a maximum value, ρ, or to estimate the outside variable from the behavior pattern, put $\bar{A} = 0$ and $\bar{\alpha} = 0$, because ρ is invariant under a shift of origin. The score of each subcategory can be determined by solving

$$\left(\frac{\partial \rho}{\partial X_{sk}} \right) = 0 \qquad (s = 1, 2, \ldots, R; \; k = 1, 2, \ldots, K_s). \tag{A.3}$$

Appendix II

Design of Electroacoustic Systems

II.1. The IACC of a Two-Channel-Loudspeaker-Reproduction System

The IACC of multiple-channel-loudspeaker-reproduction systems has been discussed for the case of wide-band noise of 250 Hz to 2 kHz (Damaske and Ando, 1972). Since a two-loudspeaker-reproduction system is the fundamental one that is often used, the optimal two loudspeaker directions for music signals and noise are discussed here, in which the IACC is minimized in a listening room (Ando, 1978).

Figure AII.1 demonstrates the calculated values of the IACC by Equation (3.25) for the symmetric loudspeaker system as a function of the horizontal angles $|\xi|$ ($\eta = 0°$), for example, with Music Motif A. The geometric loudspeaker arrangement is illustrated in the upper part of this figure. In this calculation, data of the measured interaural cross-correlation for a single directional sound were used (Ando, 1985). Optimal horizontal angles showing minima of the IACC (Ando, 1978) may be found at around:

$$24°, 60°, 135°, \text{ and } 150°.$$

These optimal angles for several other music motifs and for wide-band noise are listed in Table AII.1. Clearly, the common angles showing significant minima for all of the sound sources may be observed near two angles

$$|\xi| \approx 26° \text{ and } 151°.$$

In order to realize a smaller value of the IACC in an actual sound field, additional loudspeakers and/or reflectors in the listening room can be taken into consideration.

II.2. A System of Controlling Temporal Factors

In order to obtain the preferred condition of the temporal criteria for the existing sound field in a room, a real-time-control system, involving both the initial time-

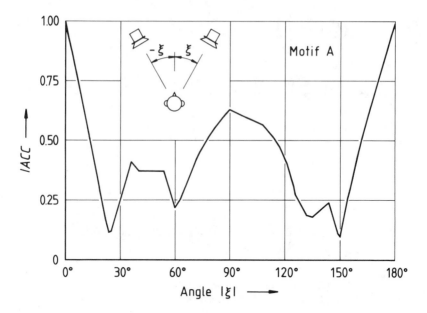

FIGURE AII.1. The value of the IACC for the two-channel-loudspeaker-reproduction sys-
tem, as a function of the horizontal angle $|\xi|(\eta = 0°)$ to the listener. The correlation values
needed for calculation of the IACC are listed in Tables D.1 and D.2 of the reference (Ando,
1985).

TABLE AII.1.   Horizontal angles to a listener indicating the IACC minima
for the standard two-channel loudspeaker reproduction system.

Sound source	Horizontal angle		$\mid \xi \mid (\eta = 0°)$		[Degrees]		
Music A	—	24	—	60	135	150	—
Music B	12	27	45	67	133	152	166
Music C	—	27	45	63	135	151	—
Music D	12	27		63	130	151	—
Noise*	—	26		63	126	151	—
Total	—	26 ± 2	—	63 ± 3	132 ± 4	151 ± 1	—

* Bandpass filtered noise, 0.25–2.0 kHz (Ando, 1985).

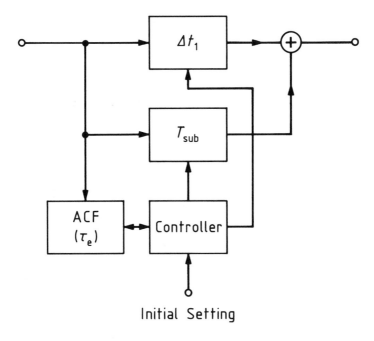

Initial Setting

FIGURE AII.2. A real-time system for controlling temporal factors.

delay gap and the subsequent reverberation time may be designed. Control is achieved by calculating the values of τ_e of the source signal, so that the preferred delay time and reverberation may be adjusted automatically, as shown in Figure AII.2. In this operation, a fast ACF-calculation is needed to produce the initial time-delay gap between the direct sound and the first reflection and the subsequent reverberation. The calculation of the short-time moving ACF may be performed by adding only operations, for example, converting the direct sound into the three-value signals of 1, 0, and -1, such that,

$$c[p(t)] = \begin{cases} 1, & p(t) > \Delta A, \\ 0, & p(t) < \Delta A, \\ -1, & p(t) < -\Delta A, \end{cases} \qquad (A.4)$$

where ΔA is a threshold value to be set smaller than the maximum value of $p(t)$ and greater than the residual noise level of the acoustic system. The initial setting is made from the program source (Figure 7.7).

Appendix III

Time-Variant Sound Fields: Variable-Delay Time of a Single Reflection

The statistical results of the measured sound-pressure level of pure tones (Ueda and Ando, 1997) suggest that the fluctuation of the sound-pressure level is caused by a change in the time delay of reflections due to fluctuation in the path difference. Strictly speaking, any sound field is more or less a time-variant system. This phenomenon is particularly significant in a large space. In order to understand the effects of such a fluctuation on subjective attributes, an initial investigation of the just noticeable difference (JND) in a variable delay time of a single reflection is discussed here (Ueda, Furuichi, and Ando, 1997). The final goal of this investigation is to find more preferred conditions in a time-variant sound field than those in the time-invariant sound field. The experiments were conducted at the preferred time delays of a single reflection which are mentioned in Section 4.1, 120 ms and 40 ms with Music Motifs A and B, respectively.

The fluctuation interval of the delay time Δ was modulated by a sinusoidal signal with frequencies of 0.2 Hz, 0.4 Hz, 0.8 Hz, and 1.6 Hz for Music Motif A and 0.4 Hz, 0.8 Hz, 1.6 Hz, and 3.2 Hz for Music Motif B. The amplitude of the single reflection is fixed to be the same as that of the direct sound.

Results of the JND of Δ as a function of its modulation frequency, M_f, are shown in Figure AIII.1. The JND is increased by decreasing the modulation frequency for both Music Motifs A and B. Also, the same is true for the standard deviation of the JND. In other words, when the modulation frequency of the system is low enough, then the fluctuation of the delay time Δ becomes inaudible. This tendency is much more significant when the music has a fast tempo such as Music Motif B ($\tau_e = 43$ ms). For slow-tempo music, such as Music Motif A (τ_e is 127 ms), the system is sensitive to rapid changes (high value of M_f), and the JND of Δ becomes small. An approximate and tentative formula for the JND of Δ was obtained within the range of the experimental condition, and is given by

$$\Delta(\text{JND}) \approx \frac{k'}{\log \tau_e + k'' \log M_f}, \tag{A.5}$$

where $k' = 7.0$ and $k'' = 2.5$. The threshold value of $\Delta(\text{JND})$, without noticing the difference, for example, is that if the value of τ_e becomes small, then M_f may have a high value.

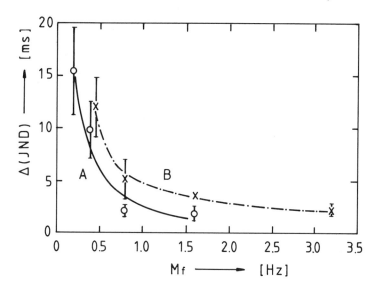

FIGURE AIII.1.

It is assumed that the subjective preference of the time-variant sound field is improved by adjusting the values of $[\Delta t_1]_p$ and $[T_{sub}]_p$ are defined for the time-invariant sound field. Initial preference judgments for sound fields with the modulated delay time of the single reflection were conducted (Atagi, Ando, and Ueda, 1998). The modulation frequency M_f was set at 0.1 Hz, a hardly sensitive condition, as shown in Figure AIII.1. The fluctuation intervals of the delay time Δ were fixed at 24 ms (Motif A) and 30 ms (Motif B), respectively. Results show that the preferred delay times of single reflection $[\Delta t_1]_p$ are shortened as 119 ms (Motif A) and 34 ms (Motif B). These may be caused by the fact that the maximum delay time of single reflection with fluctuation is $\Delta t_1 + \Delta/2$.

References

Alrutz, H. (1981). Ein neuer Algorithmus zur Auswertung von Messungen mit Pseudorauschsignalen. *Fortschritte der Akustik, DAGA'81*, 525–528.

Ando, Y., and Hattori, H. (1973). Statistical studies on the effects of intense noise during human fetal life. *J. Sound Vibration*, **27**, 101–110.

Ando, Y., Shidara, S., Maekawa, Z., and Kido, K. (1973). Some basic studies on the acoustic design of room by computer. *J. Acoust. Soc. Jpn.*, **29**, 151–159 (in Japanese with English abstract).

Ando, Y., Shidara, S., and Maekawa, Z. (1974). Simulation of sound propagation with boundary and subjective test. *Proc. 8th Intern. Congr. Acoust.*, London, p. 611.

Ando, Y., Nakane, Y., and Egawa, J. (1975). Effects of aircraft noise on the mental work of pupils. *J. Sound Vibration*, **43**, 683–691.

Ando, Y., and Kato, K. (1976). Calculations on the sound reflection from periodically uneven surfaces of arbitrary profile. *Acustica*, **35**, 321–329.

Ando, Y., and Kageyama, K. (1977). Subjective preference of sound with a single early reflection. *Acustica*, **37**, 111–117.

Ando, Y., and Hattori, H. (1977a). Effects of noise on sleep of babies. *J. Acoust. Soc. Am.*, **62**, 199–204.

Ando. Y., and Hattori, H. (1977b). Effects of noise on human placental lactogen (HPL) levels in maternal plasma. *Brit. J. Obstet. Gynaecol.*, **84**, 115–118.

Ando, Y. (1977). Subjective preference in relation to objective parameters of music sound fields with a single echo. *J. Acoust. Soc. Am.*, **62**, 1436–1441.

Ando, Y. (1978). Subjectively optimal conditions of sound fields for recording and reproducing. *96th Meeting Acoust. Soc. Am.*, joint with Acoust. Soc. Jpn. (Invited paper).

Ando, Y., and Gottlob, D. (1979). Effects of early multiple reflections on subjective preference judgments of music sound fields. *J. Acoust. Soc. Am.*, **65**, 524–527.

Ando, Y., and Imamura, M. (1979). Subjective preference tests for sound fields in concert halls simulated by the aid of a computer. *J. Sound Vibration*, **65**, 229–239.

Ando, Y., and Morioka, K. (1981). Effects of the listening level and the magnitude of the interaural crosscorrelation (IACC) on subjective preference judgment of sound field. *J. Acoust. Soc. Jpn.*, **37**, 613–618 (in Japanese with English abstract).

Ando, Y., and Alrutz, H. (1982). Perception of coloration in sound fields in relation to the autocorrelation function. *J. Acoust. Soc. Am.*, **71**, 616–618.

Ando, Y., Okura, M., and Yuasa, K. (1982). On the preferred reverberation time in auditoriums. *Acustica*, **50**, 134–141.

Ando, Y., Takaishi, M., and Tada, K. (1982). Calculations of the sound transmission over theater seats and methods for its improvement in the low-frequency range. *J. Acoust. Soc. Am.*, **72**, 443–448.

Ando, Y., Otera, K., and Hamana, Y. (1983). Experiments on the universality of the most preferred reverberation time for sound fields in auditoriums. *J. Acoust. Soc. Jpn.*, **39**, 89–95 (in Japanese with English abstract).

Ando, Y. (1983). Calculation of subjective preference at each seat in a concert hall. *J. Acoust. Soc. Am.*, **74**, 873–887.

Ando, Y. (1985). *Concert Hall Acoustics.* Springer-Verlag, Heidelberg.

Ando, Y. (1986). Physical properties of sound in rooms and subjective effects in man. *Proc. 12th Intern. Congr. Acoust.*, Toronto, Plenary 2 (Invited paper).

Ando, Y. (1986). Applications of the theory of subjective preference. *Proc. 12th Intern. Congr. Acoust.*, Toronto, Paper E4–14.

Ando, Y., and Kurihara, Y. (1986). Nonlinear response in evaluating the subjective diffuseness of sound field. *J. Acoust. Soc. Am.*, **80**, 833–836.

Ando, Y., Kang, S.H., and Nagamatsu, H. (1987). On the auditory-evoked potential in relation to the IACC of sound field. *J. Acoust. Soc. Jpn.* (E), **8**, 183–190.

Ando, Y., Kang, S.H., and Morita, K. (1987). On the relationship between auditory-evoked potential and subjective preference for sound field. *J. Acoust. Soc. Jpn.* (E), **8**, 197–204.

Ando, Y., and Kang, S.H. (1987). A study on the differential effects of sound stimuli on performing left- and right-hemispheric tasks. *Acustica*, **64**, 110–116.

Ando, Y. (1988). Effects of daily noise on fetuses and cerebral hemisphere specialization in children. *J. Sound Vibration*, **127**, 411–417.

Ando, Y., and Sakamoto, M. (1988). Superposition of geometries of surface for desired directional reflections in a concert hall. *J. Acoust. Soc. Am.*, **84**, 1734–1740.

Ando, Y., Okano, T., and Takezoe, Y. (1989). The running autocorrelation function of different music signals relating to preferred temporal parameters of sound fields. *J. Acoust. Soc. Am.*, **86**, 644–649.

Ando, Y., Watanabe, T., and Yamamoto, A. (1990). Effects of illuminance on subjective preference judgments for sound fields. *Reports of the Architectural Institute Japan*, Kinki Chapter, 77–80.

Ando, Y., Yamamoto, K., Nagamatu, H., and Kang, S.H. (1991). Auditory brain-stem response (ABR) in relation to the horizontal angle of sound incidence. *Acoustic Lett.*, **15**, 57–64.

Ando, Y. (1992). Evoked potentials relating to the subjective preference of sound fields. *Acustica*, **76**, 292–296.

Ando, Y., Yamamoto, M., Mitsumune, S., Yamamoto, I., and Mori, Y. (1992). A design study of a concert hall in Kobe minimizing the IACC and maximizing subjective preference. *Proc. 14th Intern. Congr. Acoust.*, Beijing, Paper F4–3.

Ando, Y., and Singh, P.K. (1994). Individual differences in subjective evaluations for sound field. *Current Topics Acoust. Res.* (India), **1**, 219–229

Ando, Y., and Setoguchi, H. (1995). Nuovi sviluppi nell'acoustica delle sale da concerto: Dati di preferenza individuale per la scelta del posto. *Atti del XXIII Congresso Nazionale*, Bologna, 11–18 (in Italian).

Ando, Y., and Chen, C. (1996). On the analysis of autocorrelation function of α-waves on the left and right cerebral hemispheres in relation to the delay time of single sound reflection. *J. Archit. Plan. Environ. Eng.*, Architectural Institute of Japan, **488**, 67–73.

Ando, Y., and Singh, P.K. (1996). A simple method of calculating individual subjective responses by paired-comparison tests. *Mem. Grad. School Sci. Technol.*, Kobe Univ., 14–A, 57–66.

Ando, Y., Johnson, B.P., and Bosworth, T. (1996). Theory of planning physical environments incorporating spatial and temporal values. *Mem. Grad. School Sci. Technol.*, Kobe Univ., 14-A, 67–92.

Ando, Y., Sato, S., Nakajima, T., and Sakurai, M. (1997). Acoustic design of a concert hall applying the theory of subjective preference, and the acoustic measurement after construction. *Acustica Acta Acustica*, 83, 635–643.

Ando, Y., and Singh, P.K. (1997). Global subjective evaluations for design of sound fields and individual subjective preference for seat selection. *Music and Concert Hall Acoustics, Conference Proceedings of MCHA 1995*. (Eds. Ando, Y., and Noson, D.). Academic Press, London, Chap. 4.

Atagi, J., Ando, Y., and Ueda, Y. (1998). Effects of the modulated delay-time of the single reflection on subjective preference. *Proc. 16th Intern. Congr. Acoust.*, Seattle (in print).

Barron, M. (1971). The subjective effects of first reflections in concert halls—The need for lateral reflections. *J. Sound Vibration*, **15**, 475–494.

Barron, M. (1981). Spatial impression due to early lateral reflections in concert halls: The derivation of a physical measure. *J. Sound Vibration*, **77**, 211–232.

Békésy, G. (1934). Ueber die Hoersamkeit von Konzert und Rundfunksaelen. *Eleckrische Nachrichten Technik*, **11**, 369–375.

Békésy, G. (1960). *Experiments in Hearing* (transl. and ed. Wever, E.G.). McGraw-Hill, New York.

Békésy, G. (1967). *Sensory Inhibition*. Princeton University Press, Princeton.

Beranek, L.L. (1957). Revised criteria for noise in building. *Noise Control*, **3**, 19–27.

Beranek, L.L. (1962). *Music, Acoustics and Architecture*. Wiley, Inc., New York.

Beranek, L.L. (1996). *Concert and Opera Halls: How They Sound*. Acoustical Society of America, by American Institute of Physics.

Berger, E.H. (1981). Re-examination of the low-frequency (50–1000 Hz) normal threshold of hearing in free and diffuse sound fields. *J. Acoust. Soc. Am.*, **70**, 1635–1645.

Bolt, R.H., and Doak, P.E. (1950). A tentative criterion for the short-term transient response of auditoriums. *J. Acoust. Soc. Am.*, **22**, 507–509.

Born, M., and Wolf, E. (1970). *Principles of Optics*, 4th ed., Pergamon Press, Oxford.

Bosse, G. (1997). The concert hall: The heart of music and cultural life of the concert. *Music and Concert Hall Acoustics, Conference Proceedings of MCHA 1995* (Eds. Ando, Y., and Noson, D.). Academic Press, London, Chap. 37.

Bosworth, T.L. (1997). Architecture as light sound and time. *Music and Concert Hall Acoustics, Conference Proceedings of MCHA 1995* (Eds. Ando, Y., and Noson, D.). Academic Press, London, Chap. 3.

Botte, M.C., Bujas, Z., and Chocholle, R. (1975). Comparison between the growth of averaged electroencepharic response and direct loudness estimations. *J. Acoust. Soc. Am.*, **58**, 208–213.

Buchwald, J.A.S., and Huang, C.M. (1975). Far-field acoustic response: Origins in the cat. *Science*, **189**, 382–384.

Burd, A.N. (1969). Nachhallfreir Musik fuer akustische Modelluntersuchungen. Rundfunktechn. *Mitteilungen*, **13**, 200–201.

Chen, C., and Ando, Y. (1996). On the relationship between the autocorrelation function of the α-waves on the left- and right-cerebral hemispheres and subjective preference for the reverberation time of music sound field. *J. Archit. Plan. Environ. Eng.*, Architectural Institute of Japan (AIJ), **489**, 73–80.

Chen, C., Ryugo, H., and Ando, Y. (1997). Relationship between subjective preference and the autocorrelation function of left and right cortical α-waves responding to the noise-burst tempo. *J. Archit. Plan. Environ. Eng.*, Architectural Institute of Japan (AIJ), **497**, 67–74.

Chernyak, R.I., and Dubrovsky, N.A. (1968). Pattern of the noise images and the binaural summation of loudness for the different interaural correlation of noise. *Proc. 6th Intern. Congr. Acoust.*, Tokyo, Paper A-3–12.

Cocchi, A., Farina, A., and Rocco, L. (1990). Reliability of scale-model research: A concert hall case. *Appl. Acoust.*, **30**, 1–13.

Dai, G.H., and Ando, Y. (1983). Generalized analysis of sound scattering by diffusing walls. *Acustica*, **53**, 296–301.

Damaske, P. (1967/68). Subjektive Untersuchungen von Schallfeldern. *Acustica*, **19**, 199–213.

Damaske, P., and Ando, Y. (1972). Interaural crosscorrelation for multichannel loudspeaker reproduction. *Acustica*, **27**, 232–238.

Davies, W.D.T. (1966). Generation and properties of maximum-length sequences. Parts 1–3, *Control*, **10**, 302–433.

Davis, A.E., and Wada, J.A. (1974). Hemispheric asymmetry: Frequency analysis of visual and auditory evoked responses to non-verbal stimuli. *Electroencephalography and Clinical Neurophysiology*, **37**, 1–9.

Dubrovskii, N.A., and Chernyak, R.I. (1969). Binaural loudness summation under varying degrees of noise correlation. *Soviet Phys. Acoust.*, **14**, 326–332.

Edward, R.M. (1974). A subjective assessment of concert hall acoustics. *Acustica*, **30**, 183–195.

Fraisse, P. (1982). Rhythm and tempo. *In the Psychology of Music.* (Ed. Deutsch, D.). Ed., Academic Press, Orlando, Fl. Chapter 6.

Frampton, K. (1985). *Modern Architecture: Critical History.* Thames and Hudson, New York.

Frampton, K. (1987). *GA Architect 8, Tadao Ando* (Ed. Futagawa, Y.). A.D.A. Edita, Tokyo.

Fujiwara, K. (1997). Sound reflection and absorption of a QR-type Schroeder diffuser. Music and Concert Hall Acoustics, *Conference Proceedings of MCHA 1995* (Eds. Ando, Y., and Noson, D.). Academic Press, London, Chapter 21.

Galin, D., and Ellis, R.R. (1975). Asymmetry in evoked potentials as an index of lateralized cognitive processes: Relation to EEG alpha asymmetry. *Neuropsychologia*, **13**, 45–50.

Gueth, W. (1987). Der Wolfton. *Acustica*, **63**, 35–41.

Haas, H. (1951). Ueber den Einfluss eines Einfachechos auf die Hoersamkeit von Sprache. *Acustica*, **1**, 49–58.

Hayashi, C. (1952). On the prediction of phenomena from qualitative data and the quantification of qualitative data from the mathematico-statistical point of view. *Ann. Inst. Statist. Math.*, III, 69–98.

Hayashi, C. (1954a). Multidimensional quantification I. *Proc. Japan Acad.*, **30**, 61–65.

Hayashi, C. (1954b). Multidimensional quantification II. *Proc. Japan Acad.*, **30**, 165–169.

Hecox, K., and Galambos, R. (1974). Brain stem auditory evoked responses in human infants and adults. *Arch. Otolaryngology*, **99**, 30–33.

Henry, J. (1857). On acoustics applied to public buildings. *Annual Report of the Board of Regents of the Smithsonian Institution*, Washington, DC, 54, 221–234.

Hidaka, T. (1996). Personal communication.

Hidaka, T., Beranek, L.L., and Okano, T. (1995). Interaural cross-correlation, lateral fraction, and low- and high-frequency sound levels as measures of acoustical quality in concert halls. *J. Acoust. Soc. Am.*, **98**, 988–1007.

Houtgast, T., Steeneken, H.J.M., and Plomp, R. (1980). Predicting speech intelligibility in rooms from the modulation transfer function. I. General room acoustics. *Acustica*, **46**, 60–72.

Ikeda, Y. (1997). Designing a contemporary classic concert hall using computer graphics. *Music and Concert Hall Acoustics, Conference Proceedings of MCHA 1995* (Eds. Ando, Y., and Noson, D.). Academic Press, London, Chap. 2.

Jasper, H.H. (1958). Report of the committee on method of clinical examination in electroencephalography, 1957 (Appendix). The ten–twenty electrode system of the international federation. *Electroencephalogr. Clin. Neurophysiol*, **10**, 371–375.

Jewett, D.L. (1970). Volume-conducted potentials in response to auditory stimuli as detected by averaging in the cat. *Electroenceph. Clin. Neurophysiol.*, **28**, 609–618.

Johnson, R., Kahle, E., and Essert, R. (1997). Variable coupled volume for music performance. *Music and Concert Hall Acoustics, Conference Proceedings of MCHA 1995* (Eds. Ando, Y., and Noson, D.). Academic Press, London, Chap. 38.

Jordan, V.L. (1969). Acoustical criteria for auditoriums and their relation to model techniques. *J. Acoust. Soc. Am.*, **47**, 408–412.

Kang, S.H., and Ando, Y. (1985). Comparison between subjective preference judgments for sound fields by different nations. *Mem. Grad. School Sci. Technol.*, Kobe Univ., 3–A, 71–76.

Katsuki, Y., Sumi, T., Uchiyama, H., and Watanabe, T. (1958). Electric responses of auditory neurons in cat to sound stimulation. *J. Neurophysiol.*, **21**, 569–588.

Keet, M.V. (1968). The influence of early lateral reflections on the spatial impression. *Proc. 6th Intern. Congr. Acoust.*, Tokyo, Paper E-2–4.

Kiang, N.Y.-S. (1965). *Discharge Pattern of Single Fibers in the Cat's Auditory Nerve*. MIT Press, Cambridge, MA.

Kimura, D. (1973). The asymmetry of the human brain. *Sci. Amer.* **228**, 70–78.

Knudsen, V.O. (1929). The hearing of speech in auditorium. *J. Acoust. Soc. Am.*, **1**, 56–82.

Kobayashi, Y., Tokuhiro, H., Owaki, M., Okuno, K., Yamada, S., and Ando, Y. (1997). Acoustical design and characteristics of the atrium for music performance in a hotel. *Music and Concert Hall Acoustics, Conference Proceedings of MCHA 1995* (Eds. Ando, Y., and Noson, D.). Academic Press, London, Chap. 27.

Korenaga, Y., and Ando, Y. (1993). A sound-field simulation system and its application to a seat-selection system. *J. Audio Eng. Soc.*, **41**, 920–930.

Kripke, D.F. (1972). An ultradian biologic rhythm associated with perceptual deprivation and REM sleep. *Psychosomatic Medicine*, **34**, 221–234.

Kuno, S., Ohno, H., and Nakahara, N. (1987). A Two-dimensional model expressing thermal sensation in transitional conditions. *ASHRAE Trans.*, **93**, 396–406.

Kuttruff, H. (1991). *Room Acoustics*, 3rd edn. Elsevier Applied Science, London.

Lauterborn, W., and Parlitz, U. (1988). Methods of chaos physics and their application to acoustics. *J. Acoust. Soc. Am.*, **84**, 1975–1993.

Lev, A., and Sohmer, H. (1972). Sources of averaged neural responses recorded in animal and human subjects during cochlear audiometry (Electro-cochleogram). *Arch. Klin. Exp. Ohr., Nas.-u. Kehlk. Heilk.*, **201**, 79–90.

Levy, J., and Trevarthen, C. (1976). Metacontrol of hemispheric function in human split-brain patients. *J. Exper. Psychol.: Human Perception and Performance*, **2**, 299–312.

Lynch, K. (1972). *What Time Is This Place?* MIT Press, Cambridge, MA.

MacNair, W.A. (1930). Optimum reverberation time for auditoriums. *J. Acoust. Soc. Am.*, **1**, 242–248.

Maki, F. (1994a). Kirishima International Concert Hall. *Shinkenchiku*, **69** (11), 115–136 (in Japanese).

Maki and Associates. (1994b). Works by Fumihiko Maki. *GA Japan—Environmental Design* (11), **1994**, 200–215 (in Japanese).

Maki, F. (1997). Sound and figure: Concert hall design. *Music and Concert Hall Acoustics, Conference Proceedings of MCHA 1995* (Eds. Ando, Y., and Noson, D.). Academic Press, London, Chap. 1.

Mandelbrot, B.B. (1982). *The Fractal Geometry of Nature*. W.H. Freeman, San Francisco, CA.

Marshall, A.H. (1968a). Acoustical determinants for the architectural design of concert halls. *Archit. Sci. Rev. (Australia)*, **11**, 81–87.

Marshall, A.H. (1968b). Concert hall shapes for minimum masking of lateral reflections. *Proc. 6th Intern. Congr. Acoust.*, Tokyo, E-49–52 (Paper E-2-3).

Masuda, K., and Fujiwara, K. (1997). Sound reflection from periodical uneven surfaces. *Music and Concert Hall Acoustics, Conference Proceedings of MCHA 1995* (Eds. Ando, Y., and Noson, D.). Academic Press, London, Chap. 19.

Mehrgardt, S., and Mellert, V. (1977). Transformation characteristics of the external human ear. *J. Acoust. Soc. Am.*, **61**, 1567–1576.

Merthayasa, I Gde N., Hemmi, H., and Ando, Y. (1994). Loudness of a 1 kHz pure tone and sharply (1080 dB/Oct.) filtered noises centered on its frequency. *Mem. Grad. School Sci. Technol.*, Kobe Univ., 12-A, 147–156.

Merthayasa, I Gde N., Ando, Y., and Nagatani, K. (1995). The effect of interaural cross correlation (IACC) on loudness in a free field. *Proc. Seminar National and Pameran Akustik '95* (Indonesia), 174–184.

Merthayasa, I Gde N., and Ando, Y. (1996). Variation in the autocorrelation function of narrow band noises: Their effect on loudness judgment. *Japan and Sweden Symposium on Medical Effects of Noise*.

Merthayasa, I Gde N., and Ando, Y. (1997). The autocorrelation function of sound at each seat in a hall. *Music and Concert Hall Acoustics, Conference Proceedings of MCHA 1995*, (Eds. Ando, Y., and Noson, D.). Academic Press, London, Chap. 26.

Meyer, J. (1995). Influence of communication on stage on the musical quality. *Proc. 15th. Intern. Congr. Acoust.*, Trondheim, 573–576.

Morimoto, M., Shunto, M., and Maekawa, Z. (1982). Effect of arriving direction of echo on echo-disturbance of speech. *Trans. Environ. Eng.*, Architectural Institute of Japan, **4**, 65–70 (in Japanese).

Morse, P.M. (1948). *Vibration and Sound*. McGraw-Hill, New York.

Mosteller, F. (1951). Remarks on the method of paired comparisons. III. *Psychometrika*, **16**, 207–218.

Mouri, K., Mori, Y., and Ando, Y. (1996). On the subjective preference of sound field in a courtyard for listeners. (unpublished).

Mouri, K., and Ando, Y. (1998). Relationship between subjective preference for sound fields and brain activity with vocal music. *Proc. 16th Intern. Congr. Acoust.*, Seattle (in print).

Mueller, G., and Lauterborn, W. (1996). The bowed string as a nonlinear dynamical system. *ACUSTICA Acta Acoustica*, **82**, 657–664.

Nagamatsu, H. Kasai, H., and Ando, Y. (1989). Relation between auditory evoked potential and subjective estimation—Effect of sensation level. *Report of Meeting of Acoust. Soc. Jpn.*, 391–392 (in Japanese).

Nakajima, T. and Ando, Y. (1991). Effects of a single reflection with varied horizontal angle and time delay on speech intelligibility. *J. Acoust. Soc. Am.*, **90**, 3173–3179.

Nakajima, T. (1992). Speech intelligibility and clarity related to spatial-binaural factor for sound field in a room. PhD Thesis at Graduate School of Science and Technology, Kobe University.

Nakajima, T., Ando, Y., and Fujita, K. (1992). Lateral low-frequency components of reflected sound from a canopy complex comprising triangular plates in concert halls. *J. Acoust. Soc. Am.*, **92**, 1443–1451.

Nakajima, T., Yoshida, J., and Ando, Y. (1993). A simple method of calculating the interaural cross-correlation function for a sound field. *J. Acoust. Soc. Am.*, **93**, 885–891.

Nakajima, T., and Ando, Y. (1997). Calculation and measurement of acoustic factors at each seat in the Kirishima International Concert Hall. *Music and Concert Hall Acoustics, Conference Proceedings of MCHA 1995* (Eds. Ando, Y. and Noson, D.). Academic Press, London, Chap. 5.

Nakayama, I. (1984). Preferred time delay of a single reflection for performers. *Acustica*, **54**, 217–221.

Nakayama, I. (1986). Preferred delay conditions of early reflections for performers. *Proc. Vancouver Symposium Acoust. and Theatre Planning for Performing Arts*, 27–32.

Nakayama, I., and Uehara, T. (1988). Preferred direction of a single reflection for a performer. *Acustica*, **65**, 205–208.

Nakayama, I. (1988). Acoustic conditions preferred by performers. *Second Joint Meeting of Acoust. Soc. Am. and Acoust. Soc. Jpn.*, Honolulu.

Nishio, K., and Ando, Y. (1996). On the relationship between the autocorrelation function of the continuous brain waves and the subjective preference of sound field in change of the IACC. *J. Acoust. Soc. Am.*, **100** (A), 2787.

Ocker, C. (1995). Personal communication.

Ohgushi, K., and Sakuma, M. (1997). Conveyance of a player's intentions in performances of percussion—interaction between auditory and visual information. *Music and Concert Hall Acoustics, Conference Proceedings of MCHA 1995* (Eds. Ando, Y., and Noson, D.). Academic Press, London, Chap. 40.

Onchi, Y. (1961). Mechanism of the middle ear. *J. Acoust. Soc. Am.*, **33**, 794–805.

Onitsuka, H., and Kawakami, F. (1997). Numerical study of energy dissipation in QR diffusers. *Music and Concert Hall Acoustics, Conference Proceedings of MCHA 1995* (Eds. Ando, Y., and Noson, D.). Academic Press, London, Chap. 20.

Ono, K., and Ando, Y. (1996). A study on loudness of sound field in relation to the reverberation time. *Reports of Architectural Institute of Japan*, Kinki chapter, 121–124 (in Japanese).

Othmer, E., Hayden, M.P., and Segelbaum, R. (1969). Encephalic cycles during sleep and wakefulness in humans: A 24-hour pattern. *Science*, **164**, 447–449.

Pickles, J.O. (1982). *An Introduction to the Physiology of Hearing*. Academic Press, London.

Puria, S., Rosowski, J.J., and Peake, W.T. (1993). Middle-ear pressure gain in humans: Preliminary results. *Proceedings of the International Symposium on Biophysics of Hair Cell Sensory Systems*, (Eds. Duifhuis, H., Horst, J.W., Dijk, P., and Netten, S.M). 345–351. World Scientific, Singapore.

Rasmussen, A.T. (1943). *Outline of Neuro-Anatomy*, 3rd edn. William C. Brown, Dubuque, IA.

Rindel, J.H. (1986). Attenuation of sound reflections due to diffraction. *Nordic Acoustical Meeting*, 20–22.

Roederer, J.G. (1995). Are magnetic storms hazardous to your health? *Eos. Trans. Amer. Geophys. Union*, **76**, 444–445.

Rubinstrin, M., Feldman, B., Fischler, H., and Frei, E.H. (1966). Measurement of stapedial-footplate displacements during transmission of sound through the middle ear. *J. Acoust. Soc. Am.*, **40**, 1420–1426.

Sabine, W.C. (1900). Reverberation. The American Architect and the Engineering Record (Sabine, W.C. Prefaced by Beranek, L.L. (1992). *Collected Papers on Acoustics*. Peninsula, Los Altos, CA, Chap. 1.

Sabine, W.C. (1912). Architectural acoustics: The correction of acoustical difficulties. *Architectural Quarterly of Harvard University*, 3–23.

Sakai, H., Sato, S., and Ando, Y. (1996). Effects of multiple scattering by trees in a forest as a music-performance space. *J. Acoust. Soc. Am.*, **100** (A), 2838.

Sakai, H., Singh, P.K., and Ando, Y. (1997). Inter-individual differences in subjective preference judgments of sound fields. *Music and Concert Hall Acoustics, Conference Proceedings of MCHA 1995* (Eds. Ando, Y., and Noson, D.). Academic Press, London, Chap. 13.

Sakai, H., Sato, S., and Ando, Y. (1998). Orthogonal acoustical factors of sound fields in a forest compared with those in a concert hall. *J. Acoust. Soc. Am.* (to be published).

Sakurai, M., Korenaga, Y., and Ando, Y. (1997). A sound simulation system for seat selection. *Music and Concert Hall Acoustics, Conference Proceedings of MCHA 1995* (Eds. Ando, Y., and Noson, D.). Academic Press, London, Chap. 6.

Sakurai, M., Aizawa, S., and Ando, Y. (1998). A diagnostic system measuring orthogonal factors of sound fields in a scale model of concert hall. *Proc. 16th Intern. Congr. Acoust.*, Seattle (in print).

Sato, S., and Ando, Y. (1996). Effects of interaural crosscorrelation function on subjective attributes. *J. Acoust. Soc. Am.*, **100** (A), 2592.

Sato, S., and Ando, Y. (1997). On the apparent source width for music sources related to the IACC and the width of the interaural crosscorrelation function (W_{IACC}). *J. Acoust. Soc. Am.*, **102** (A), 3188.

Sato, S., Mori, Y., and Ando, Y. (1997). On the subjective evaluation of source locations on the stage by listeners. *Music and Concert Hall Acoustics, Confer-*

ence Proceedings of MCHA 1995 (Eds. Ando, Y., and Noson, D.). Academic Press, London, Chap. 12.

Sato, S., Ota, S., and Ando, Y. (1998). Individual preference on the delay time of a single reflection for cellists. *Proc. 16th Intern. Congr. Acoust.*, Seattle (in print).

Schroeder, M.R. (1962). Natural sounding artificial reverberation. *J. Audio Eng. Soc.*, **10**, 219–223.

Schroeder, M.R. (1965a). New method of measuring reverberation time. *J. Acoust. Soc. Am.*, **37**, 409–412.

Schroeder, M.R. (1965b). Response to *comments on "New method of measuring reverberation time"* [Smith, P.W., *J. Acoust. Soc. Am.* **38**, 359(L) (1965)]. *J. Acoust. Soc. Am.*, **38**, 359–361.

Schroeder, M.R., Gottlob, D., and Siebrasse, K.F. (1974). Comparative study of European concert halls: Correlation of subjective preference with geometric and acoustic parameters. *J. Acoust. Soc. Am.*, **56**, 1195–1201.

Schroeder, M.R. (1979). Binaural dissimilarity and optimum ceilings for concert halls: More lateral sound diffusion. *J. Acoust. Soc. Am.*, **65**, 958–963.

Seraphim, H.P. (1961). Ueber die Wahrnehmbarkeit mehrerer Rueckwuerfe von Sprachshall. *Acustica*, **11**, 80–91.

Shaw, E.A.G. (1974). Transformation of sound pressure level from the free field to the eardrum in the horizontal plane. *J. Acoust. Soc. Am.*, **56**, 1848–1861.

Shoda, T., and Ando, Y. (1996). Calculation of speech intelligibility using an autocorrelation function. *J. Acoust. Soc. Am.*, **100** (A), 2819.

Shoda, T., and Ando, Y. (1998). Calculation of speech intelligibility using four orthogonal factors extracted from the autocorrelation function of source and sound field signals. *Proc. 16th Intern. Congr. Acoust.*, Seattle (in print).

Singh, P.K., Ando, Y., and Kurihara, Y. (1994). Individual subjective diffuseness responses of filtered noise sound fields. *Acustica*, **80**, 471–477.

Singh, P.K., and Ando, Y. (1996). Individual difference in subjective preference judgments of sound field (unpublished).

Sperry, R.W. (1974). Lateral specialization in the surgically separated hemispheres. *The Neurosciences: Third study program* (Eds. Schmitt, F.O., and Worden, F.C.). MIT Press, Cambridge, MA, Chap. 1.

Strube, H. W. (1980). Scattering of a plane wave by a Schroeder diffusor: A mode-matching approach. *J. Acoust. Soc. Am.* **67**, 453–469.

Strube, H. W. (1981). More on the diffraction theory of Schroeder diffusors. *J. Acoust. Soc. Am.*, **70**, 633–635.

Sumioka, T., and Ando, Y. (1996). On the pitch identification of the complex tone by the autocorrelation function (ACF) model. *J. Acoust. Soc. Am.*, **100** (A), 2720.

Taguti, T., and Ando, Y. (1997). Characteristics of the short-term autocorrelation function of sound signals in piano performances. *Music and Concert Hall Acoustics, Conference Proceedings of MCHA 1995* (Eds. Ando, Y., and Noson, D.). Academic Press, London, Chap. 23.

Takatsu, A., Mori, Y., and Ando, Y. (1997). The architectural and acoustic design of a circular event hall in the Kobe Fashion Plaza. *Music and Concert Hall Acoustics, Conference Proceedings of MCHA1995* (Eds. Ando, Y., and Noson, D.). Academic Press, London, Chap. 30.

Takatsu, A., Hase, S., Sakai, H., Sato, S., and Ando, Y. (1998). Acoustic design and measurement of a circular hall improving the subjective preference at each seat. *Proc. 16th Intern. Congr. Acoust.*, Seattle (in print).

Takeuchi, M., Mori, Y., and Ando, Y. (1997). Sound reinforcement system of the Kirishima International Concert Hall. *Music and Concert Hall Acoustics, Conference Proceedings of MCHA 1995* (Eds. Ando, Y., and Noson, D.). Academic Press, London, Chap. 7.

Thompson, A.M., and Thompson, G.C. (1988). Neural connections identified with PHA-L anterograde and HRP retrograde tract-tracing techniques. *J. Neurosci. Meth.*, **25**, 13–17.

Thurstone, L.L. (1927). A law of comparative judgment. *Psychol. Rev.*, **34**, 273–289.

Tonndorf, J., and Khanna, S.M. (1972). Tympanic-membrane vibrations in human cadaver ears studied by time-averaged holography. *J. Acoust. Soc. Am.*, **52**, 1221–1233.

Torgerson, W.S. (1958). *Theory and Methods of Scaling*. Wiley, New York.

Ueda, Y., Furuichi, H. and Ando, Y. (1997). The just noticeable difference in variable delay time of the single reflection. *Music and Concert Hall Acoustics, Conference Proceedings of MCHA 1995* (Eds. Ando, Y., and Noson, D.). Academic Press, London, Chap. 14.

Ueda, Y., and Ando, Y. (1997). Effects of air conditioning on sound propagation in a large space. *J. Acoust. Soc. Am.*, **102**, 2771–2775.

Venekalasen, P.S., and Christoff, J.P. (1964). Seattle Opera House—Acoustical design. *J. Acoust. Soc. Am.*, **36**, 903–910.

Vitruvius (ca. 25 B.C.). *De Architecture*, Liber V, Cap. VIII (de locis consonantibus ad theatra eligendis.). See also *The Ten Books on Architecture* (Trans. Morgan, M.H. (1960)). Dover, New York.

West, J.E. (1966). Possible subjective significance of the ratio of height to width of concert halls. *J. Acoust. Soc. Am.* (A), **40**, 1245.

Wiener, F.M., and Ross, D.A. (1946). The pressure distribution in the auditory canal in a progressive sound field. *J. Acoust. Soc. Am.*, **18**, 401–408.

Wightman, F.L., (1973). Pitch and stimulus fine structure. *J. Acoust. Soc. Am.* **54**, 397–406.

Wilkens, H. (1977). Mehrdimensionale Beschreibung subjecktiver Beurteilungen der Akustik von Konzertsaelen. *Acustica*, **38**, 10–23.

Yamaguchi, K. (1972). Multivariate analysis of subjective physical measures of hall acoustics. *J. Acoust. Soc. Am.*, **52**, 1271–1279.

Zwicker, E., Flottorp, G., and Stevens, S.S. (1957). Critical band width in loudness summation. *J. Acoust. Soc. Am.*, **29**, 548–557.

Zwislocki, J.J. (1976). *The Acoustic Middle Ear Function*. Acoustic impedance and admittance—The measurement of middle ear function (Ed. Feldman, A.S., and Wilber, L.A.). Williams and Wilkins, Baltimore, MD, Chap. 4.

General Reading

Ando, Y. (1985). *Concert Hall Acoustics*. Springer-Verlag, Heidelberg.

Ando, Y., and Noson, D., Eds. (1997). *Music and Concert Hall Acoustics, Conference Proceedings of MCHA 1995*. Academic Press, London.

Békésy, G. (1967). *Sensory Inhibition*. Princeton University Press, Princeton.

Beranek, L.L. (1996). *Concert and Opera Halls. How They Sound*. Acoustical Society of America, by American Institute of Physics.

Blauert, J. (1983). *Spatial Hearing*. MIT Press, Cambridge, MA.

Cremer, L., and Mueller, H.A. (1978). *Principles and Applications of Room Acoustics*, Vols. 1 and 2 (Trans. Schultz, T.J.). Applied Science, London.

Evans, E.F., and Wilson, J.P., Eds. (1977). *Psychophysics and Physiology of Hearing*. Academic Press, London.

Jordan, V. L. (1980). *Acoustical Design of Concert Halls and Theaters*. Applied Science, London.

Katsuki, Y. (1982). *Receptive Mechanisms of Sound in the Ear*. Cambridge University Press, Cambridge.

Kuttruff, H. (1991). *Room Acoustics*, 3rd edn. Elsevier Applied Science, London.

Meyer, J. (1980). *Acoustics and the Performance of Music*. Verlag das Musikinstrument, Frankfurt.

Morse, P.M. (1948). *Vibration and Sound*. McGraw-Hill, New York.

Pickles, J.O. (1982). *An Introduction to the Physiology of Hearing*. Academic Press, London.

Roederer, J.G. (1995). *The Physics and Psychophysics of Music, An Introduction*. 3rd edn., Springer-Verlag, New York.

Sabine, W.C. (Prefaced by Beranek, L.L.) (1992). *Collected Papers on Acoustics*. Peninsula, Los Altos, CA.

Schroeder, M.R. (1984). *Number Theory in Science and Communication*. Springer-Verlag, Heidelberg.

Schroeder, M.R. (1991). *The Fractals, Chaos, Power Laws, Minutes From an Infinite Paradise*. W.H. Freeman, San Francisco, CA.

Tohyama, M., Suzuki, H., and Ando, Y. (1995). *The Nature and Technology of Acoustic Space*. Academic Press, London.

Vitruvius (Translated by Morgan, M.H.) (1960). *The Ten Books on Architecture*. Dover, New York.

Yost, W.A., and Gourevitch, G. (1987). *Directional Hearing*. Springer-Verlag, New York.

Zwicker, E., and Fastl, H. (1990). *Psychoacoustics, Facts and Models*. Springer-Verlag, Heidelberg.

Glossary of Symbols

(The number in parentheses signifies the relevant equation.)

A	Total amplitude of reflections, (4.6).
$A(P - N)$	Amplitude of the slow vertex response (SVR).
A_n	Pressure amplitude of the nth reflection determined by the $(1/r)$ law, A_0 being unity, (3.15).
c	Speed of sound in air [m/s], (3.16).
c	Constant to determine the preferred initial time-delay gap between the direct sound and the first reflection, (6.2) which is applied for any subjective responses.
d	Percentage of violation, (9.9).
d_n	Distance between the source and the observation point for each reflection, $n = 1, 2, 3, \ldots$ See (3.15).
f	Frequency [Hz].
$f_{l,r}(t)$	Sound pressures at the left and right ear entrances, (3.14).
$f_{l,r}(t; R_0)$	Visual image at the left and right retina, (12.5).
F	Number of sound fields, (9.6), (9.9).
g	Function of any subjective responses in relation to physical factors, (4.8), (12.1).
$g_{l,r}(x)$	Scale values of subjective preference in relation to the left- and right-hemispheric factors, (12.2)–(12.4).
$g_{l,r}(t)$	Impulse responses of a room, (3.13).
$g(x_i)$	Scale values of orthogonal factors, $i = 1, 2, 3, 4$, (4.9).
$h_{l,r}(t)$	Impulse responses between the sound source and the left and right ear-canal entrances in a free field, (3.13); Impulse responses between the light source and each position of the left and right retina, (12.5).
IACC	Magnitude of the interaural cross-correlation function, (2.3), (3.26). See also Figure 3.7.
IALD	Interaural level difference, Section 3.4.2.
IATD	Interaural time difference, Section 3.4.2.
k	Constant to determine preferred initial time-delay gap

	between the direct sound and the first reflection, (6.2) which is applied for any subjective responses.
L_{eq}	Equivalent sound level, (3.8).
N_w	Period of diffuser. See Figure 8.7.
N_m	Latencies at the mth maxima of the SVR, $m = 1, 2, 3$.
p	Significance level.
$p(t)$	Source signal, (3.2), (3.3).
P	Physiological magnitude corresponding to the IACC, (5.2).
$P(i > j)$	Probability that i is preferred to j, (9.6).
$P(\omega)$	Fourier Transform of $p(t)$, (3.2).
$P_d(\omega)$	Power density spectrum, (3.1).
P_m	Latencies at the mth minima of SVR, $m = 1, 2, 3$.
r	Correlation coefficient.
r_0	Position of the sound source, $r_0 = (x_0, y_0, z_0)$.
R_0	Position of the light source, $R_0 = (x_0, y_0, z_0)$, (12.5).
s	$=$ IACC, (6.5).
s	Integer.
$s(t)$	Impulse response of the A-weighting filter corresponding to ear sensitivity; $f'_{l,r}(t) = f_{l,r}(t) * s(t)$ in (3.22).
S	Total scale value of preference, (4.9) or the total scale value of a single subjective response, (12.1).
S	Total surface in a room, (3.20).
SCC	Short-term cross-correlation coefficient. See Table 2.1.
S_i	Scale values of preference as a function of the listening level, initial time delay gap between the direct sound and the first reflection, subsequent reverberation time, and IACC, respectively, $i = 1, 2, 3, 4$, (4.9) and (4.10).
SD	Scale of dimension of a concert hall, $\Delta t_1 = 22$ (SD), (4.2).
SI	Speech intelligibility [%], (6.5).
t	$=$ STI, (6.5).
t	Time [s].
T	Time interval [s].
T	Time factor of reflector, (12.5).
T_{60}	Reverberation time defined by Sabine, (2.2).
T_{sub}	Subsequent reverberation time defined by the decay rate to decrease to 60 dB just after early reflections [s].
$[T_{sub}]_p$	Calculated preferred subsequent reverberation time [s], (4.7).
$w_n(t)$	Impulse response describing reflection properties of boundaries, $n = 1, 2, 3, \ldots , (3.13), (12.5)$.
W_{IACC}	Width of the IACC or width of the interaural cross-correlation function at the τ_{IACC}, as defined in Figure 3.7.
Z_{ab}	Scale value obtained by the probability, (9.2).
$\bar{\alpha}$	Averaged absorption coefficient, (3.20), (3.21).

α_i	Weights of the scale value of preference for each orthogonal factor, $i = 1, 2, 3, 4$, (4.10).
$\delta(t)$	Dirac delta function.
Δt_1	$= (d_1 - d_0)/c$ [s]; Initial time-delay gap between the direct sound and the first reflection.
$\Delta t_{1'}$	The most preferred delay time of the first reflection for alto-recorder soloists, (7.1).
Δt_n	Delay time of the nth reflections relative to the direct sound [s], (3.17).
$[\Delta t_1]_p$	Calculated preferred initial time-delay gap between the direct sound and the first reflection [s], (4.1), (4.3), (4.5); see also the equation between (4.6) and (4.7).
$\Delta\omega$	$= 2\pi(f_2 - f_1)$, (3.28), (6.3).
$\Delta\omega_c$	$= 2\pi(f_2 + f_1)$, (3.28), (6.3).
$\phi_{lr}(t)$	Normalized interaural cross-correlation function, (3.23).
$\phi_p(t)$	Normalized ACF, (3.7).
$\Phi_{ll}(0), \Phi_{rr}(0)$	ACFs at the origin of time corresponding to averaged sound energies at the left ear and the right ear, respectively, (3.23), (3.24).
$\Phi_{lr}(t)$	Interaural cross-correlation function, (3.22).
$\Phi_p(t)$	Autocorrelation function (ACF), (3.4).
η	Elevation angle to a listener.
λ	Poorness of fit for the model, (9.7).
ν	Time delay [s].
σ	Time delay [s].
$\sum$	Summation.
τ	Time delay [s].
τ_e	Effective duration of the ACF, defined by the delay time at which the envelope of the normalized ACF becomes 0.1 (the ten percentile delay) [s].
τ_{IACC}	Interaural delay time at which the IACC is defined in Figure 3.7.
τ_p	Most preferred delay time of the first reflection, $[\Delta t_1]_p = \tau_e$, (4.1).
ω	Angular frequency, $\omega = 2\pi f$ [rad/s].
ξ	Horizontal angle to a listener.

Abbreviations

ABR	Auditory brainstem response.
ACF	Autocorrelation function.
AEP	Auditory evoked potential.
ASW	Apparent source width.
BESTI	Best ear STI, (6.6).

CBW	Continuous brain wave.
EDT	Early decay time.
IACC	Interaural cross-correlation.
IALD	Interaural level difference.
IATD	Interaural time difference.
JND	Just noticiable difference.
LL	Listening level measured by dB(A).
REM	Rapid eye movement.
STI	Speech transmission index.
SVR	Slow vertex response.

Author Index

Subject Index